AFRICA

South Africa's
Gold Mines
& the Politics
of Silicosis

AFRICAN ISSUES

AFRICAN ISSUES

South Africa's Gold Mines & the Politics of Silicosis

JOCK McCULLOCH

James Currey
is an imprint of
Boydell & Brewer Ltd
PO Box 9, Woodbridge
Suffolk IP12 3DF (GB)
www.jamescurrey.com
and of
Boydell & Brewer Inc.
668 Mt Hope Avenue
Rochester, NY 14620-2731 (US)
www.boydellandbrewer.com

Published in Southern Africa
(South Africa, Namibia,
Lesotho, Swaziland & Botswana)
by Jacana Media (Pty) Ltd
10 Orange Street
Sunnyside
Auckland Park
2092
Johannesburg
South Africa

© Jock McCulloch 2012

First published 2012

1 2 3 4 5 15 14 13 12

The right of Jock McCulloch to be identified as
the author of this work has been asserted in accordance with
sections 77 and 78 of the Copyright, Designs and Patents Act 1988

This publication is printed on acid-free paper

British Library Cataloguing in Publication Data
A catalogue record for this book is available on request from the British Library

ISBN 978-1-84701-059-9 (James Currey paper)
ISBN 978-1-4314-0718-7 (Jacana paper)

The publisher has no responsibility for the continued existence or accuracy of URLs for
external or third-party internet websites referred to in this book, and does not guarantee
that any content on such websites is, or will remain, accurate or appropriate.

Papers used by Boydell & Brewer are natural, recycled products
made from wood grown from sustainable forests.

MIX
Paper from
responsible sources
FSC
www.fsc.org   FSC® C013604

Typeset in 9/11 Melior with Optima display
by Avocet Typeset, Chilton, Aylesbury, Bucks
Printed and bound in Great Britain
by CPI Group (UK) Ltd, Croydon, CR0 4YY

CONTENTS

LIST OF PHOTOGRAPHS

DEDICATION

To Rhett & Janet Kahn

Preface

South Africa in the Twentieth Century

On the eve of majority rule in 1990 the President of the Chamber of Mines K. W. Maxwell made some telling observations about South Africa. Per capita disposable income was falling and more than five million South Africans were unemployed. Half of the adult population was illiterate and half of the country's children were not attending school. South Africa had just 60,000 students in technical and higher education. With half the population Australia had over 800,000.[1] Forty years of apartheid had denied most South Africans basic civil rights: it had also stunted the national economy. Maxwell was no doubt aware that the mines which had made the country the dominant industrial power on the continent had helped to create one of the world's most unequal societies.

The legacy inherited four years later by the newly-elected ANC government was every bit as daunting as Maxwell had suggested. South Africa had massive deficits in health care, housing, education and employment. The disparities between the rich and poor were among the highest in the world and they fell along the lines of colour which had so preoccupied minority governments.

South Africa's modern history lies at the convergence of two colonial systems, the British and the Dutch. Its transformation from a rural society to an industrial state at the end of the nineteenth century was accelerated by the discovery of diamonds near Kimberly in 1867 and gold at Johannesburg in 1886. It is a story of late colonial conquest, mineral wealth and the persistence of racial ideologies which came to be embodied in apartheid. Unlike most of colonial Africa, South Africa attracted a large number of white settlers. It also differed from the new world colonies of the USA and Australia where the indigenous populations were soon outnumbered by colonists. In 1910 whites accounted for only 20 per cent of South Africa's population, a proportion which remained stable until the 1960s. Today it is just under 10 per cent. South Africa's ethnic compo-

---

[1] K. W. Maxwell, Presidential Address Chamber of Mines of South Africa 100th Annual General Meeting, 19 June 1990:5.

sition was made more complex by the importation of Indian labour in the 1860s and by the presence of a large 'Coloured' or 'mixed-race' population in the Cape.

Labour shortages are a constant theme in South Africa's colonial history. In the 1860s, the Natal sugar plantations turned to indentured Indian workers to meet their needs. After diamonds were discovered at Kimberley, white farmers were soon complaining to government about their difficulties in attracting workers. Contests over labour became intense with the opening of gold mines. A large number of drillers from Cornwall's tin and copper mines found their way to Johannesburg. To fill a shortfall between 1904 and 1906 the Rand mines imported over 60,000 indentured Chinese. Black peasant farmers were reluctant to enter the mines and their struggles with the state over access to land were always in part about the labour needs of the mines and white-owned farms.

There has been much debate among historians about the origins of segregation and apartheid. The lineaments of both can be found in the Dutch period which ended in 1806. Segregation was also a feature of British rule, especially in the years following the South African War of 1899–1902. By the time of Union in 1910, many of the elements of the divided society South Africa was to become were clearly visible. From 1910 to 1948 the dominant social model was segregation; from 1948 to 1990 it was apartheid. What distinguished the two periods was the intensity of the racial politics. One way to trace that history is through the legislation which widened the boundaries between South Africa's ethnic communities. The Mines and Works Act of 1911 legitimised a colour bar in the workplace by restricting certain types of skilled work to whites. The Natives Land Act of 1913 prohibited African land purchase outside native reserves; the Natives (Urban Areas) Act of 1923 segregated urban settlement, and the 1936 Representation of Natives Act abolished the remnant African franchise in the Cape Province. In 1948 the National Party under D. F. Malan introduced apartheid and within five years it had passed a swathe of legislation further denying political, economic and civil liberties to eighty per cent of the population.

Segregation was part of the fabric of South Africa's industrial revolution which began at the end of the nineteenth century. At times it stimulated economic growth, at others it was an impediment. The colour bar which excluded black men and women from higher-paying work was an invention of white labour not white capital. The Transvaal Chamber of Mines opposed job reservation during the passage of the Mines and Works Act of 1911 because it added to the costs of production. The bar was also a source of friction between capital and black labour. There was a massive strike by black miners in 1913 over low wages and rising prices. In February 1920 around 70,000 miners again went on strike. One of their grievances was the bar which reduced their wages by ignoring their skills.

The political struggles over white privilege in the workplace were at times intense. Following World War I the price of gold fell. To maintain profitability employers wanted to cut the number of white miners, reduce the wages of those who remained and fill the vacancies with blacks. To do so required diluting the colour bar. In protest, white miners began a strike in March 1922 which quickly turned into the Rand Revolt. Martial law was declared and the uprising was eventually crushed by the state at the cost of over 200 lives. Almost 5,000 protesters were arrested and the wages of white labour were cut dramatically.[2] Two years later, in a political backlash, the Prime Minister Jan Smuts lost the general election to a National and Labour Parties coalition. In a compromise with the white trade unions the new government introduced the 1924 Industrial Conciliation Act, and the 1925 Wage Act. They were followed by the Mines and Works Amendment Act of 1926 which reinforced the colour bar. One immediate result was to widen the gap between the wages of white and black miners, a pattern that was in contrast to other sectors of the economy where over time the differential shrunk.[3] William Gemmill of the Chamber of Mines played a pivotal role in the passage of the legislation which amounted to a re-structuring of South Africa's industrial relations system.[4]

The scale and economic dominance of the mining industry drew the state and big business together. Francis Wilson has written that in choosing their labour policies the architects of apartheid took the gold mining industry as their model.[5] The relationship was also at times fraught. The mines' need for labour was met through taxation, land reservation, and a state-sanctioned recruiting system.[6] The mines relied on the state to provide infrastructure, especially railways and electrification. In its turn the state became dependent upon mine revenue to fund health services and an education system to benefit its white constituents. To meet those needs mining taxation was far higher than in Australia or the USA.[7] Between 1933 and 1939 direct taxation on the gold industry accounted for a third of state revenue. Between 1948 and 1960 the development of the Free State mines saw state revenue from gold rise three fold. To some degree the new mines financed apartheid.

The inequalities between whites and blacks became more marked in the period after World War II as inflation eroded the wages of black miners.[8] An increasing number of workers moved into manufacturing; those who remained on the mines began organising. In August 1946 the

[2] David Yudelman, *The Emergence of Modern South Africa: State, Capital, and the Incorporation of Organised Labour on the South African Gold Fields, 1902–1939* (Cape Town: David Philip, 1984): 190.

[3] It was the beginning of a long trend. In 1921 the ratio was 15 to 1 but by 1969 it had grown to 20 to 1. Francis Wilson *Labour in the South African Gold Mines 1911–1969* (Cambridge: Cambridge University Press, 1972): 45.

[4] David Yudelman, *The Emergence of Modern South Africa*: 156.

[5] Francis Wilson, *Labour in the South African Gold Mines*:13.

[6] See Shula Marks and Richard Rathbone (eds), *Industrialisation and Social Change in South Africa: African Class Formation, Culture and Consciousness, 1870–1930* (London: Longman, 1982).

[7] Merle Lipton, *Capitalism and Apartheid* (London: Gower/Maurice Temple Smith, 1985): 259.

[8] Francis Wilson, *Labour in the South African Gold Mines*: 46.

African Mine Workers' Union went on strike demanding a minimum wage of 10s. a day. It was a massive protest involving almost 100,000 men. The smashing of that strike signalled a new phase in which the gold industry turned increasingly to labour drawn from outside of South Africa.

The introduction of apartheid in 1948 was a messy and complex process which further racialised labour markets and restricted access to education and health care for most South Africans.[9] It also increased segregation in transport, public spaces and even sexual relations. Among apartheid's major features was state control over the movement of labour. The pass-book system introduced initially in 1923 was extended. Outside designated 'homelands', black South Africans had to carry a passbook at all times, proving they were authorised to live or move in 'White' South Africa. The document, similar to an internal passport containing the bearer's fingerprints, photograph, and employer's name, became one of the most despised symbols of minority rule. Resistance to the laws led to tens of thousands of arrests each year and eventually ignited the Sharpeville massacre in March 1960 when 69 protesters were killed by police.

The sheer scale of gold mining meant that throughout the twentieth century the national economy was dominated by large corporations and none was larger than Anglo American. The opening of the Free State mines saw Anglo become the industry's leader: by 1958 its capital base exceeded that of all the other mining houses combined.[10] During the 1960s Anglo American invested heavily in local industry and within a decade the Group had major holdings in steel, ferro-alloys, printing, publishing, computers and textiles.[11] In 1983 seven companies controlled 80 per cent of the shares listed on the Johannesburg stock exchange. Anglo American and its subsidies accounted for half that total.[12]

While from the outside apartheid appeared to benefit all whites there was no natural community of white interests, and the material and political advantages which flowed from the system varied from one sector of the economy to another. They also varied over time.[13] Apartheid justified and supported the system of oscillating migrant labour upon which the mines profitability depended but it forced employers to hire expensive white workers. Apartheid provided employment in state utilities for whites with little education and few skills but that in turn inflated the size of the state sector and with it the cost. It offered comfort to those who

[9] See William Beinart, *Twentieth-Century South Africa* (Oxford: Oxford University Press, 1994): 135–162; and Deborah Posel, *The Making of Apartheid, 1948–1961: Conflict and Compromise* (Oxford: Clarendon Press,1991).

[10] Francis Wilson, *Labour in the South African Gold Mines*: 26.

[11] By 1976 Anglo's various interests were held through a network of 152 subsidiaries. Duncan Innes, *Anglo American and the Rise of Modern South Africa* (New York: Monthly Review Press, 1984): 204.

[12] Merle Lipton, *Capitalism and Apartheid*: 242.

[13] See William Beinart and Saul Dubow (eds), *Segregation and Apartheid in Twentieth-Century South Africa* (London: Routledge, 1995).

feared the growth of a large, black proletariat but finally gave them no security from the civil strife that followed. Apartheid constricted the growth of a skilled work force and stifled the domestic market for local products often against the wishes of the mining and manufacturing industries.

The one sector of the white community which benefited most from apartheid was the farmers. National Party policies gave them access to cheap labour and inflated their incomes by the use of subsidies and tariffs.[14] There was little state regulation of work conditions or wages and at law farm workers had fewer rights than miners. Until 1974, when the last of the Masters and Servants Ordinances was abolished, farm workers were unable to leave their place of employment without permission. Infractions were a criminal offence. The impact of those discriminatory labour policies was felt throughout the country.

One way to measure the impact of minority rule is in the area of public health. There were half a dozen famines between 1910 and 1960. While few South Africans died, a significant proportion of the black population experienced what Diana Wylie terms 'starving on a full stomach'. A low calorific intake or a poorly-balanced diet, meant that many black South Africans were unable to properly grow, work, bear a child or recover from illness. In addition to low wages and land scarcity, the major cause was the price of food. Marketing boards established by the Marketing Act of 1937 protected white farmers by artificially inflating food prices on the domestic market.[15] The Boards which controlled the sale of most farm produce until the 1980s made food expensive in a setting in which the incomes of black families were static. The result was widespread malnutrition.

The contradiction between the success of the national economy and the human cost of apartheid became obvious in the 1960s and 1970s. During what was a period of strong economic growth, government set out to eradicate the remaining pockets of black settlement or 'black spots' in white designated areas. That programme required the forced removal of more than three million people, most of them to homelands or Bantustans. Families were broken apart, communities destroyed and jobs lost. In 1960 the Bantustans accounted for 4.2 million people or 39 per cent of the black population: in 1980 that had grown to over 11 million or 53 per cent.[16] It was common for urban populations to be relocated hundreds of kilometres to desolate rural areas where there was no industry, no jobs, few schools and no health services. The film *The Last Grave at Dimbaza* which was smuggled out of South Africa in 1974 showed the human costs of forced removals in a bleak town in the Eastern Cape, a region which

---

[14] Merle Lipton, *Capitalism and Apartheid*: 80.
[15] Diana Wylie, *Starving on a Full Stomach: Hunger and the Triumph of Cultural Racism in Modern South Africa* (Charlottesville and London: University Press of Virginia, 2001): 80.
[16] William Beinart and Saul Dubow, 'Introduction: The Historiography of Segregation and Apartheid' in William Beinart and Saul Dubow (eds), *Segregation and Apartheid*: 16.

was a source of labour for the gold mines. The film caused a scandal in Britain.

While the forced removals were in full swing the fortunes of the white minority and in particular the fortunes of Afrikaners were on the rise. The state sector became an Afrikaner reserve and government-funded enterprises a training ground for Afrikaner entrepreneurs. In 1964 Federale Mynbou took over General Mining Ltd, and in 1968 Tom Muller became the first Afrikaner chairman of the Chamber of Mines of South Africa, renamed thus in that year. Studies from the 1970s show that Afrikaners in state employment who benefited from the National Party's 'civilised labour policies' were the most strongly opposed to any easing of apartheid.[17]

By the mid-1970s there was growing concern among big business about the rising costs of what had become the world's most racialised society. A large part of public revenue was being squandered on the National Party's programmes and the state had become the largest single employer in the country.[18] The mining industry was particularly concerned about the political consequences. Barlow Rand and Anglo American were global corporations, and many of their trading partners were hostile to minority rule. Despite continuing state violence in the townships there was a slow revival of African trade unions.[19] National Party governments made some modest concessions and in 1979 black workers were accepted under the legal definition of employees. Five years later mixed race trade unions were allowed.

There has been much debate as to whether apartheid was a perversion of capitalism or merely an odious variation. In some respects apartheid, which began in the same year as the signing of the Universal Declaration of Human Rights, ran against the tide of post-war history. During voting on the Declaration at the United Nations there were eight abstentions and of those only two states, Saudi Arabia and South Africa, were not aligned with the Soviet bloc. The introduction of apartheid also coincided with the breaking up of Europe's empires in Africa, South-East Asia and the Pacific. In that context, apartheid appears to be a residue from the age of imperialism. On closer scrutiny South Africa's history is not so unusual. In the first half of the twentieth century fears of racial decline were common to industrial states. The eugenics movement was influential especially in the USA where Jim Crow laws also flourished in the south. The racial policies of Australia differed from apartheid in their scale but not their intensity. Australian governments which used selective immigration policies to maintain a white nation also denied Aboriginal Australians the right to vote in national elections until 1962. In some states Aboriginal stockmen, miners and domestic workers were paid in hand-outs of sugar, flour, tea and tobacco because they had no legal right to receive wages.

[17] Merle Lipton, *Capitalism and Apartheid*: 212.
[18] Merle Lipton, *Capitalism and Apartheid*: 235.
[19] Merle Lipton, *Capitalism and Apartheid*: 340.

Children of so-called mixed race were routinely removed from their black mothers to erase their Aboriginality. Until 1967 indigenous Australians were not counted in the national population census.

Given the weight of its internal contradictions it is surprising that apartheid lasted so long. Its end in 1990 was brought about by a number of factors. Resistance by black South Africans in the townships which was played out nightly to TV audiences in Britain and the USA made the country virtually ungovernable. The increasing levels of state violence further eroded the legitimacy of minority rule. There was also pressure from the frontline states of Zimbabwe, Angola, and Mozambique for political change. They withdrew labour from South Africa's mines and farms and provided a base for insurgents. Opposition from the international community hardened as sporting boycotts were bolstered by economic sanctions. Finally, structural changes within South Africa's mining and manufacturing industries made apartheid unsustainable.

The chronic conflict between the state and most of South Africa's citizens forms the background for the story of silicosis on the gold mines.

ACKNOWLEDGMENTS

Like all books, this one was written with the help of many people. They include Tony Davies, David Rees, Jill Murray and Mary Ross at the National Institute of Occupational Health in Johannesburg, and Jonny Meyers at the University of Cape Town. A special debt is owed to Anna Trapido for her path-breaking research in the Eastern Cape which made this project possible. I was fortunate to interview Oluf Martiny, who spent his career working as a medical officer with the Witwatersrand Native Labour Association in Johannesburg. I also owe thanks to Emile Schepers and his late father Gerrit Schepers in Grand Falls, Virginia who provided me with a home for a week. Gerrit talked candidly about his career in mine medicine in Johannesburg from the late 1940s. I owe special thanks to Ntja Moletsane, Zakaria Mofokeng Mokaeane, Oscar Dlamini, and Khanhelo Mofokeng Mokaeane who have spent their working lives on the gold mines of the Free State. Aline Tsoene, whose husband Bennett Tsoene is a miner, spoke with me at length about how families adjust to the burden of chronic illness.

I am also indebted to a consortium of researchers consisting of Paul-André Rosental, Alberto Baldasseroni, Francesco Carnevale, David Rosner, Jerry Markowitz, Joe Melling, Eric Geerkens, Martin Lengwiler, and Julia Moses. The consortium, of which I am a member, has been working on a joint history of silicosis. Thanks are due to Richard Spoor in Nelspruit and Richard Meeran in London who, against the odds, have been fighting the legal battle on behalf of miners and their families. I owe thanks to the staff of the South African National Archives, the Archives at the University of Johannesburg, The National Archives of Malawi, The National Archives of Swaziland, the National Archives of Botswana and the British National Archives. I also must thank Dalene Hosch at the Library of Parliament in Cape Town and Rochelle Keene at the Adler Museum of Medicine, Faculty of Health Sciences, University of the Witwatersrand. The Museum has generously given permission for the use of photographs which appear in this book.

Regular conversations in Melbourne with Beris Penrose who is writing the first history of silicosis in Australia have been a joy. So too has a correspondence with Geoff Tweedale in Manchester with whom I have worked on the history of occupational disease and in particular on the history of asbestos mining and manufacture.

David Rosner and Dunbar Moodie read the manuscript and made a number of valuable suggestions which I have followed. Jillian Smith has been a research assistant throughout this long project: her help has been invaluable.

I wish also to thank Jaqueline Mitchell at James Currey who has supported this book. Finally I am indebted to Barbara Metzger who edited the manuscript and to the Australian Research Council which provided funding for the project.

Jock McCulloch
North Fitzroy, Australia

LIST OF ABBREVIATIONS

Many of the key organisations changed their names over the period covered in this history. The International Labour Office became the International Labour Organisation while the Miners' Phthisis Medical Bureau, which was established in 1916, underwent numerous transformations from the Silicosis Medical Bureau to the Pneumoconiosis Research Unit to its present reincarnation as the National Institute for Occupational Health. In 1977 the Native Recruiting Corporation and the Witwatersrand Native Labour Association were amalgamated to form The Employment Bureau of Africa or Teba.

The Witwatersrand Chamber of Mines was founded in 1889. In 1897 it became the Chamber of Mines of the South African Republic until 1902, when it became the Transvaal Chamber of Mines. In 1953 the name was again changed to the Transvaal and Orange Free State Chamber of Mines, until 1968 when it became the Chamber of Mines of South Africa.

A number of the colonial territories which supplied labour to the gold mines changed their names at independence. Bechuanaland became Botswana, Basutoland became Lesotho, Southern Rhodesia became Zimbabwe, Northern Rhodesia became Zambia and Nyasaland became Malawi.

The British currency of the time, pounds, shillings and pence (£sd) were in use in South Africa until 1961 (pegged in value to that of the United Kingdom until 1931). There were twenty shillings to the pound and twelve pennies to the shilling. The units were abbreviated as, e.g., £10. 5s. 6d. (ten pounds, five shillings and six pence). Further, two pounds, ten shillings (for example) was often further abbreviated to £2/10/- and two shillings and six pence to 2/6. The new currency, the Rand, is sub-divided into 100 cents, expressed as, e.g., R6.50.

CCOD   Compensation Commission for Occupational Diseases
COIDA   Compensation for Occupational Injuries and Diseases Act No. 130 of 1993

DNA	Department of Native Affairs
DOTS	Directly Observed Therapy Short-course
GMO	Group Medical Officers Committee
GPC	Gold Producers Committee
ILO	International Labour Organisation
PRU	Pneumoconiosis Research Unit
MBOD	Medical Bureau for Occupational Diseases
MMOA	Mine Medical Officers' Association
MPMB	Miners' Phthisis Medical Bureau
MMR	Mass Miniature Radiography
MWU	Mine Workers Union
NIOH	National Institute for Occupational Health
NRC	Native Recruiting Corporation
NUM	National Union of Mineworkers
ODMWA	Occupational Diseases in Mines and Works Act No. 78 of 1973
PDC	Pulmonary Disability Committee
PRU	Pneumoconiosis Research Unit
SAIMR	South African Institute of Medical Research
SIMRAC	The Safety in Mines Research Advisory Committee
SMB	Silicosis Medical Bureau
Teba	The Employment Bureau of Africa
WHO	World Health Organisation
WNLA	Witwatersrand Native Labour Association

A BRIEF CHRONOLOGY

1867	Discovery of diamonds near Kimberly.
1886	Gold mining begins on the Witwatersrand.
1899–1902	South African (Boer) War.
1903	Weldon Commission Report into miners' phthisis.
1910	Union of South Africa proclaimed.
1911	The first Miners' Phthisis Act. Silicosis in gold miners becomes a compensatible disease.
1912	Foundation of the South African Native National Congress, (later the ANC).
1913	Native Land Act.
1913	Strike by black miners over low wages.
1914	National Party founded.
1916	Creation of the Miners' Phthisis Medical Bureau. Tuberculosis in gold miners becomes a compensatible disease.
1920	Strike by black miners over the colour bar and low wages.
1921	Mine Medical Officers Association is formed.
1922	Rand Revolt.
1923	Native Urban Areas Act. The pass laws introduced.
1930	ILO Silicosis Conference in Johannesburg.
1936	The Native Representation Act extinguishes the limited franchise in the Cape.
1938	The ILO Pneumoconiosis Conference, Geneva.
1946	African Mine Workers Union strike.
1948	D. F. Malan's National Party wins the general elections. Apartheid begins.
1950–53	Major apartheid legislation passed.
1956	Pneumoconiosis Research Unit is established.
1959	Second Johannesburg Conference on Pneumoconiosis.
1960	Sharpeville massacre.
1961	South Africa becomes a Republic. The ANC is banned.
1973	Strikes. An independent trade union movement is

	launched.
1976	Students protests and repression in Soweto. The Transkei becomes the first independent Bantustan or Homeland.
1976	Erasmus Commission into Occupational Health.
1984–86	Mass protests. State of emergency declared.
1985	COSATU trade union federation is founded.
1986	Mass protests. Pass laws are rescinded.
1990	Nelson Mandela released from goal.
1994	First majority elections held. ANC government elected.
1995	Leon Commission Report into health and safety on the mines.
1995	ILO/WHO Global Programme for the Elimination of Silicosis begins.
2006	Test case against the gold industry for compensation launched in Johannesburg.

1. Group of entrants for the mining industry assemble at the WNLA
compound, Johannesburg. Time 6am. Undated, early 1950s
(*Source:* Museum Africa: Johannesburg)

1
Gold Mining
& Life-Threatening
Disease

In October 2006 the case of Thembekile Mankayi versus AngloGold Ashanti, a subsidiary of Anglo American, was filed in the Johannesburg High Court. Mr Mankayi had for sixteen years worked as a gold miner at the Vaal Reefs Mine and then been dismissed in 1995 because of ill health. Suffering from silicosis and tuberculosis, for which he was seeking compensation, Mankayi claimed that AngloGold knew or should have known that silica dust causes serious lung disease and that it failed to protect him from the risk of injury. It had long been assumed that the various Mines Acts stretching back to 1911 precluded litigation, and Mankayi had first had to establish his right to sue his former employer.[1] His claim was dismissed by the High Court and then by the Supreme Court of Appeal. Finally, in March 2011, Mankayi was given leave by the Constitutional Court to have his case heard.[2] Tragically, he died a week before the ruling that found in his favour. In spite of his death, the decision is likely to precipitate a class action involving hundreds of thousands of miners from South Africa, Lesotho, Malawi, Swaziland, Mozambique and Botswana, all of which supplied labour to Anglo American mines. The industry's defence is that it has been as much the victim of failed science as Thembekile Mankayi – that the mining houses, which include some of the world's most powerful corporations, had no knowledge of any risk to their employees. It is the same defence that has been used with diminishing success by asbestos companies in USA, UK, and Australian courts.[3]

Anyone familiar with the official history of silicosis or miners' phthisis in South Africa will find the recent developments in Johannesburg

[1] See Thembekile Mankayi versus AngloGold Ashanti Ltd., High Court of South Africa (Witwatersrand Local Division) Case No. 06/22312 2. The parallel case of Mankayi Mbini has stalled on the grounds of jurisdiction. See Mankayi Mbini versus Anglo American Corporation of South Africa Ltd., High Court of South Africa (Witwatersrand Local Division) Case No. 04/18272, 2 August 2004.
[2] See the Matter between Thembekile Mankayi Applicant and AngloGold Ashanti Ltd., Respondent, Constitutional Court of South Africa Case CCT 40/10 [2011] ZACC 3, Constitutional Court of S.A. Judgment 3 March 2011.

surprising. In the period from 1902 to 1912 South Africa's gold mines did face a silicosis crisis. The host ore on the Rand has a high silica content, and the use of pneumatic drills and gelignite produced clouds of fine dust that could destroy a miner's lungs in a few years. From 1910 a series of laws on miners' phthisis and the mines saw the South African industry invest heavily in dust suppression technologies and medical surveillance. In 1912 South Africa became the first state to compensate silicosis as an occupational disease. Four years later gold miners with pulmonary tuberculosis, a recognised sequela of silicosis, became eligible for awards. The introduction of ventilation, blasting regulations and watering down reduced the palpable dust and with it the incidence and severity of silicosis. From the 1920s minority governments in Pretoria were extolling the success of an important industry in preventing a life-threatening disease.

It is not surprising that the Rand mines had an enviable reputation. The data presented in the annual reports of the Miners' Phthisis Medical Boards was one of ceaseless improvement in mine safety in which South Africa led the world. In the period from 1917 to 1920, for example, the reported silicosis rate among white miners was 2.195 per cent,[4] and by 1935 it had fallen to 0.885 per cent. The rate for black miners was even lower, 0.129 per cent in 1926–27 and 0.122 per cent in 1934–35.[5] These data were accepted as authoritative by the international community and South Africa's achievements featured prominently in science and policy debates in Australia, France, Belgium, Germany, Italy, Japan, the UK and the USA. In 1930 the British specialist Arthur Hall referred to Johannesburg as 'the mecca for silicosis researchers' and praised South Africa for leading the world in safety, medical care and compensation.[6] South Africa's admirers had, however, little understanding of working conditions on the Rand or of how the official data were collected.

The end of minority rule brought a revised understanding of South African science. In 1995 the Commission of Inquiry into Safety and Health in the Mining Industry, the first inquiry of its kind to draw upon the testimony of black miners, found that dust levels were hazardous and had probably been so for more than fifty years.[7] Subsequent research by Anna Trapido, David Rees, Tony Davies and Jill Murray at the National Institute of Occupational Health (NIOH) in Johannesburg and by Neil White, T. W. Steen, Rodney Erlich and Jonny Myers at the University of Cape Town

---

[3] See Jock McCulloch and Geoffrey Tweedale, *Defending the Indefensible: The Global Asbestos Industry and its Fight for Survival* (Oxford: Oxford University Press, 2008).

[4] See 'Table 1, Incidence of Silicosis and Tuberculosis' in *The Prevention of Silicosis on the Mines of the Witwatersrand* (Pretoria: Government Printer, 1937): 236.

[5] *The Prevention of Silicosis*, 242.

[6] Arthur J. Hall, 'Some Impressions of the International Conference on Silicosis' *The Lancet*, 30 September 1930: 657–58.

[7] *Report of the Commission of Inquiry into Safety and Health in the Mining Industry* (Pretoria: Department of Minerals and Energy Affairs, 1995): 51–53.

identified a pandemic of hitherto undiagnosed and uncompensated disease. These studies put the silicosis rate in living miners at between 22 and 30 per cent, more than a hundred times higher than the official rate between 1930 and 1990.[8] Jill Murray's post-mortem research estimated that up to 60 per cent of miners will eventually develop silicosis.[9] The previous underreporting raises the spectre of a backlog of compensation claims amounting to what one British finance analyst estimates at US$100 billion.[10]

The 1995 Commission of Inquiry into Safety and Health in the Mining Industry (or Leon Commission) concluded that tens of thousands of miners had contracted silicosis without the disease being diagnosed and without receiving the compensation to which they were entitled. That finding gave impetus to research indicating that in 2000 as many as one hundred and ninety-six thousand miners in South Africa and a further eighty-four thousand in neighbouring states were entitled to awards.[11] Gavin Churchyard's research from 2003 suggested that more than eighty thousand men then currently employed should receive compensation.[12] Since then the numbers have risen. The most recent estimates, from June 2011, put the potential claimants at three hundred thousand. [13] Those figures have two components: current miners, many of whom will develop lung disease, and men who left the industry without receiving benefits, many of whom had been retrenched during the 1980s. In August 2009 the Chamber of Mines took the Compensation Commissioner for Occupational Disease to the Gauteng High Court to contest a proposed thirtyfold increase in the levy for the Compensation Fund, from R0.32 a risk shift to R9.38.[14] The rise was to cover the backlog of claims and the impact of HIV/AIDS, which greatly increases the chances of a miner's contracting tuberculosis. In March 2011 the Chamber approached the High Court for a declaratory order against the Compensation Fund, arguing that because

[8] T. W. Steen, K. M. Gyi, N. W. White, et al., 'Prevalence of Occupational Lung Diseases among Botswana Men Formerly Employed in the South African Mining Industry' *Occupational and Environmental Medicine*, no. 54 (1997): 19–26; B. Girdler-Brown et al., 'The Burden of Silicosis, Pulmonary Tuberculosis, and COPD among Former Basotho Goldminers' *American Journal of Industrial Medicine* 51 (2008): 640–47; Jill Murray, Tony Davies and David Rees, 'Occupational Lung Disease in the South African Mining Industry: Research and Policy Implementation' *Journal of Public Health Reports* 32 (2011): 65–79.

[9] Jill Murray, 'Development of Radiological and Autopsy Silicosis in a Cohort of Gold Miners Followed Up in Retirement' paper presented at the Research Forum, National Institute for Occupational Health, Johannesburg, 26 May 2005.

[10] Mike Cohen and Carli Lourens, 'Blowback from the Apartheid Era' *Bloomberg Businessweek,* 6 June–12 July, 2011; see also Anna Trapido, 'An Analysis of the Burden of Occupational Lung Disease in a Random Sample of Former Gold Mineworkers in the Libode District of the Eastern Cape' Ph.D. diss., University of the Witwatersrand, 2000: 196–202.

[11] Trapido, 'The Burden of Occupational Lung Disease': 196–202.

[12] G. Churchyard et al., 'Silicosis Prevalence and Exposure-Response Relationship in Older Black Mineworkers on a South African Gold Mines Safety' in *Mines Research Advisory Committee* (SimHealth, 15 May 2003).

[13] Cohen and Lourens, 'Blowback from the Apartheid Era'.

[14] 'Mining Industry Fights Levy Hike' *Business Report*, 26 August 2009.

most of the claims come from former miners the government rather than the industry should pay.[15]

In addition to silicosis, the mines have played a major role in the spread of tuberculosis. According to current estimates the incidence of tuberculosis in South Africa is among the highest in the world, and the tuberculosis rate among miners is ten times higher than for the general population.[16] Key studies of miners from Botswana and Lesotho show elevated rates of infection. Radiological evidence from Libode in the Eastern Cape puts the incidence as high as 47 per cent.[17] David Stuckler and his colleagues have found that, even allowing for HIV/AIDS, miners in sub-Saharan Africa and especially those in South Africa's gold mines have a higher rate of tuberculosis than any working population in the world.[18] That many miners are migrant workers leads to the transmission of disease to rural communities and across national borders.[19] There is evidence that South Africa's mines play a role in spreading tuberculosis to the general population similar to that of prisons in the former Soviet Union.[20]

The new science raises many questions. Among the most important is how the current disease rates can be reconciled with the earlier official data. There is no evidence that working conditions have deteriorated dramatically since the beginning of majority rule, and while diagnostic methods have improved over time they have not been revolutionised. The mine workforce has become more stable, but labour stabilisation has long been a feature of South Africa's mines and cannot explain the discrepancy between current and past disease rates. Rather than any dramatic increase in silicosis and tuberculosis, it seems likely that the rates have been very high for decades. If that is so, then the history of these diseases needs rewriting.

Silicosis is one of the oldest of the occupational diseases, with a biomedical literature stretching back into the 1830s. The early science came from doctors in mining or manufacturing towns like Manchester, Glasgow and Edinburgh in Britain and Kalgoorlie, Ballarat and Bendigo in Australia. Because they treated large numbers of hard rock miners, stonemasons or cutlery grinders, local physicians were the first to notice an association between lung disease and particular industries. In his history

---

[15] 'Chamber of Mines Goes to Court over Levy Hike' *Miningmx*, 15 March 2011. www.miningmx.com/news/markets (accessed 1 May 2012).

[16] Jaine Roberts, *The Hidden Epidemic amongst Former Miners: Silicosis, Tuberculosis and the Occupational Diseases in Mines and Works Act in the Eastern Cape, South Africa* (Westville: Health Systems Trust, 2009): 46.

[17] Roberts, *The Hidden Epidemic*: 51.

[18] David Stuckler, Sanjay Basu, Martin McKee and Mark Lurie, 'Mining and Risk of Tuberculosis in Sub-Saharan Africa' *American Journal of Public Health* 101 (2010): 524.

[19] Around 40 per cent of the adult male tuberculosis patients in Lesotho's hospitals have worked in South African mines. *The Mining Sector: Tuberculosis and Migrant Labour in South Africa* (Johannesburg: AIDS and Rights Alliance for Southern Africa, 2008): 2.

[20] Stuckler et al., 'Mining and Risk of Tuberculosis': 529.

of working-class life in Manchester, first published in 1844, Frederick Engels writes at length of 'grinders' rot' or what is now known as silicosis. Grinders would often begin work at the age of fourteen, and it usually took six years before the first symptoms appeared. Engels quotes a Sheffield physician on what happened to the men who made tools, cutlery, locks or nails and were exposed to silica dust at the grinding wheels:

> They suffer from shortness of breath at the slightest effort in going up hill or up stairs, they habitually raise the shoulders to relieve the permanent and increasing want of breath; they bend forward and seem in general to feel most comfortable in the crouching position in which they work. Their complexion becomes dirty yellow, their features express anxiety, they complain of pressure upon the chest.

These men would cough, have night sweats, lose weight and invariably die from consumption (tuberculosis). With the appearance of the first symptoms a return to the grinding wheel meant certain death. This description is almost identical to the one composed by the South African Commission of Inquiry into Miners' Phthisis and Pulmonary Tuberculosis in 1912.[21]

In terms of the number of cases, the number of deaths and the cost of litigation, silicosis has a good claim to being the paradigmatic occupational disease of industrialism. The emergence of silicosis as a major hazard during the nineteenth century coincided with the introduction of power tools and especially pneumatic drills and gelignite in the hard rock mines of Bendigo and Kalgoorlie in Australia, the Tri-State region of the USA and Johannesburg in South Africa. The inhalation of free silica dust can lead to scarring or fibrosis of the lungs. The classic symptoms are breathlessness or 'air hunger', a persistent cough and weight loss.[22] Silicosis has a long latency period, and especially in its early stages it is not easy to diagnose. It will progress even after a worker has left a dusty occupation. By itself silicosis can be fatal, and it also greatly increases an individual's chances of contracting pulmonary tuberculosis. This has often been observed in Southern Africa, where the rural communities from which migrant labour was drawn had had little prior exposure to infection.

Silicosis is a chronic disease, and while the symptoms can nowadays be alleviated, the process of decline cannot. In Southern Africa the burden of illness falls on families and particularly on women. In addition to the loss of income and future earnings, other costs are borne by family members. Older children may be taken out of school either to care for the sick or to earn money. The income spent on medical treatment and trans-

---

[21] Frederick Engels, *The Condition of the Working Class in England* (London: Panther Books, 1969 [1897]): 230–38, quotation at 230; *Report of a Commission of Inquiry into Miners' Phthisis and Pulmonary Tuberculosis* (Cape Town: Government Printer, 1912): 9.

[22] See 'The Development of Silicosis,' in Elaine Katz, *The White Death: Silicosis on the Witwatersrand Gold Mines 1886–1910* (Johannesburg: Witwatersrand University Press, 1994): 215–20.

port often leads to the sale of assets and eventually to poverty.[23] The International Labour Organization (ILO) / World Health Organization (WHO) International Programme on the Global Elimination of Silicosis, launched in 1995, aims at the eventual elimination of a disease that continues to kill thousands of workers each year in both the industrial and the developing world. South Africa is a signatory to the accord.

Tuberculosis is one of the oldest and most widespread of the microparasites, and it is probable that low levels of infection have existed in Southern Africa for centuries. The disease is contracted through inhalation of airborne droplets. There will be a primary lesion in the lungs and the disease's progression will depend upon an individual's immune system. Tuberculosis is not highly infectious, but if a patient carries other diseases or is undernourished infection it is more likely to occur. From the late eighteenth century tuberculosis was the major cause of premature death in Europe and North America as crowded housing combined with malnutrition to spread infection through the new industrial cities. Effective treatment became available only after World War II with the introduction of streptomycin and then isoniazid in 1952. By that time improved housing and nutrition in the developed world had already seen the infection and mortality rates fall. The most recent major drug, rifampicin, was introduced in 1967.

Silicosis and tuberculosis affect the same parts of the lung. In a silicotic lung tubercle bacilli tend to be attracted to existing lesions, and in a tubercular lung silica particles will gravitate to areas of infection.[24] It is now understood that even without the development of silicosis exposure to silica dust will greatly increase the chances of an individual's developing tuberculosis.[25] While researchers outside South Africa preferred the term 'silicosis' to 'miners' phthisis,' in South Africa the latter term continued to be used in both the medical literature and the mines legislation until 1956, when it was replaced by 'pneumoconiosis' to bring local usage into line with international practice. A survey of safety officers in 2007 found that miners prefer the term *sifuba*, a synonym for phthisis, to describe chest disease caused by dust.[26] The survival of the term in South Africa is suggestive of a co-joined pandemic that miners themselves recognised.

[23] Roberts, *The Hidden Epidemic.*

[24] *Report of the Departmental Committee of Enquiry into the Relationship between Silicosis and Pulmonary Disability and the Relationship between Pneumoconiosis and Tuberculosis,* part 2 (Pretoria: Government Printer, 1954): 14.

[25] E. Hnizdo and J. Murray, 'Risk of Pulmonary Tuberculosis Relative to Silicosis and Exposure to Silica Dust in South African Gold Miners,' *Occupational and Environmental Medicine* 55 (1998): 496–502.

[26] *A Summary Report of Research with Mine Health and Safety Representatives: What Does Silicosis Elimination Mean to Mine Health and Safety Representatives?* D. Rees, J. Murray and F. Ingham SIM 030603 Track C, Silicosis Elimination Awareness for Persons Affected by Mining Operations in South Africa (Pretoria: Mine Health and Safety Council and National Institute of Health, 2007): 3.

Historians have long suspected the role played by the gold mines in the spread of infection. In evidence before the Leon Commission in August 1994, Francis Wilson commented: 'It seems to be generally agreed that the mining industry made a major contribution to the development of tuberculosis in South Africa amongst miners and from the miners who then circulate it back home to their families.'[27] Wilson suggested that there was probably a correlation between the intensity of migration from particular areas and the incidence of tuberculosis. In his classic study, Randall Packard presented a complex narrative in which the spread of tuberculosis in South Africa was shaped by the growth of an industrial economy based on mining. Packard identified three epidemics: the first at the beginning of the twentieth century, the second in the late 1930s, and the last coinciding with the arrival of HIV/AIDS in the 1980s. He argued that oscillating migration probably delayed the development of resistance to the disease.[28]

Dust and labour in the mines

From the middle of the nineteenth century silicosis was a common occupational disease in the USA, Australia, the UK, Italy, Germany, Japan and France. In the USA it occurred in hard rock mines, stone quarries, building construction and glass, steel and iron foundries located in dozens of states and involving thousands of individual employers. By the early 1930s all dusty industries faced problems of lung disease and litigation. In 1933 silicosis lawsuits involving more than US$30 million in claims were filed against foundries in New York State alone. Employers resolved the crisis to their advantage by having the issues of risk and compensation referred back to regional legislatures.[29] The history of silicosis in South Africa is very different. In terms of legislation, medical research and compensation, in South Africa silicosis has been a disease of one industry, the gold mines.

The Rand mines were concentrated in a single geological deposit and were distinguished by their depth, their scale, and the size of their labour force. The deposits were low-grade; it took three tons of ore to yield an ounce of gold. The host ore was brittle and abrasive, and the deposits were worked at depths of over three thousand five hundred metres. Most of the reefs were less than a metre in width, which meant that the ore had to be extracted over a wide area by labour-intensive methods.[30] Those features encouraged the dominance of a few mining houses which, because of their

[27] Francis Wilson, evidence before the Commission of Inquiry into Safety and Health in the Mining Industry, Braamfontein, Johannesburg, 1 August 1994, *Testimony* vol. 8: 746.
[28] Randall Packard, *White Plague, Black Labor: Tuberculosis and the Political Economy of Health and Disease in South Africa* (Berkeley: University of California Press, 1989).
[29] D. Rosner and G. Markowitz, *Deadly Dust: Silicosis and the Politics of Industrial Disease* (Ann Arbor: University of Michigan Press, 2nd edn, 2006): 81.
[30] *Report of the Commission of Inquiry into Safety and Health in the Mining Industry* (Pretoria: Department of Minerals and Energy Affairs, 1995): 27.

importance to employment, foreign exchange and state revenue, wielded great political influence. On the global stage the mines were just as important. In 1930 South Africa produced over 60 per cent of the gold upon which the stability of the Western financial system rested. The commercial links between South Africa's gold mines and British capital meant that knowledge of silicosis and an understanding of the means for its prevention were available before mining began in the Transvaal. That knowledge was not easily translated into the creation of safe workplaces.

Dust in mines is created by the drilling, blasting, and moving of ore. The threat to health posed by free silica is determined by particle size, mineral composition, concentration, and the duration of exposure. The host ore in South African gold mines is between 60 and 80 per cent silica, which is high by world mining standards. The discovery in 1913 by the South African chemist John McCrae that the dust particles that can penetrate lung tissue and cause silicosis are invisible to the naked eye overturned the idea that safety was proven by a lack of visible dust.[31]

Beginning in 1920, dust was measured with a konimeter, a device invented by the government mining engineer Robert Kotze in 1914, and by 1925 more than a hundred thousand samples were being taken annually.[32] The device took a snap reading lasting less than a quarter-second, which represents 1/100,000 of an eight-hour shift. Dust levels changed, however, from moment to moment, and no single moment was necessarily representative of a particular miner's experience working a particular shift. Moreover, mines were huge workplaces, and individual readings were unlikely to be typical of a mine as a whole. Beyond this, the konimeter, while reliable in recording low levels of coarse dust, was inefficient at the high levels of fine dust identified by McCrae as the most dangerous. Speaking at an international conference on silicosis in 1930, the government mining engineer Hans Pirow questioned the dust standard then being adopted by industry in the USA: 'He found with the Rand percentages [of free silica] that 300 particles per cubic centimetre could be taken as a standard but he was not convinced that it was safe.'[33] Konimetric data were convenient for setting levies but had little meaning as a measure of risk. Konimetric sampling was abandoned internationally in the 1970s but continued to be used in South Africa until 1992.

At the same time, some of the changes designed to minimise risk had unexpected consequences. The introduction of water-fed drills in 1910 reduced the palpable dust, but the water sprays increased the humidity and with it the spread of tuberculosis.[34] What never changed was the industry's dependence on cheap labour.

---

[31] John McCrae, *The Ash of Silicotic Lungs* (Johannesburg: South African Institute for Medical Research, 1918[1913]).

[32] *Silicosis: Records of the International Conference held at Johannesburg 13–27th August 1930* (London: ILO, 1930): 32.

[33] *Silicosis*: 29.

[34] Packard, *White Plague, Black Labor*: 73–91.

The mines employed a huge number of migrant workers drawn from rural communities inside and outside South Africa's borders. The workforce consisted of a unionised white labour aristocracy employed in supervisory roles and black migrants who did the bulk of the manual work. In 1910 there were 120,000 black miners and 10,000 whites; by 1929 black miners numbered 193,221 while there were 21,949 whites. As the workforce grew to over half a million in the late 1970s, that colour imbalance was maintained.[35] The size of the workforce peaked in the 1980s, and by 2010 it had shrunk to 160,000. In the first decades of mining the compounds were severely overcrowded and, in addition to silicosis, infectious diseases, especially pneumonia, tuberculosis and meningitis, were rife.[36] The mortality rate from pneumonia was shocking even by the standards of the day. According to one estimate, nearly fifty thousand black miners died during the decade to 1912.[37]

Because the price of gold was fixed, as a spokesman for the Chamber of Mines lamented in 1929, the cost of production was 'of even more vital importance in the gold mining industry than in other industries'.[38] The major costs were stores and wages. White miners were unionised, and therefore the most obvious way for employers to guarantee profitability was to reduce the wages of black miners. In addition to wages, expenditure on mine compounds, occupational health and compensation were areas in which employers could control expenditure. The Chamber readily acknowledged that South Africa's low-grade deposits could not have been worked profitably in Canada or Australia because of the high minimum wages in those countries.[39] The Rand mines were made profitable by migrant labour. From 1870 the Kimberley diamond mines established a pattern of oscillating migration whereby men from Swaziland and the Transkei who continued to live in rural areas left their families for months at a time to work on the mines. The gold mines quickly adopted that system. In 1896 the Chamber formed the Native Labour Association, later the Witwatersrand Native Labour Association (WNLA), to coordinate the supply of miners from Mozambique. A brief experiment with Chinese labour soon ended, and in 1912 the Chamber established the Native Recruiting Corporation (NRC) to organise recruiting in South Africa and the British protectorates of Basutoland (now Lesotho), Bechuanaland (now

---

[35] See Table 5, 'Employment on the Gold Mines,' in David Yudelman, *The Emergence of Modern South Africa: State, Capital, and the Incorporation of Organized Labour on the South African Gold Fields, 1902–1939* (Westport CT: Greenwood Press, 1983): 191.

[36] H. J. Simons, 'Migratory Labour, Migratory Microbes: Occupational Health in the South African Mining Industry, the Formative Years, 1870–1956,' (MS, 1960): 74.

[37] Alan H. Jeeves, *Migrant Labour in South Africa's Mining Economy: The Struggle for the Gold Mines' Labour Supply, 1890–1920* (Kingston and Montreal: McGill-Queen's University Press, 1985): 234.

[38] Statement of the Gold Producers' Committee of the Transvaal Chamber of Mines to the Miners' Phthisis Commission 1929–1930 (Johannesburg: Parliamentary Library, Cape Town, 10 December 1929): 34.

[39] Miners' Phthisis Commission Report (Cape Town: Parliamentary Library, 15 May 1931): 160.

Botswana) and Swaziland. By 1920 the NRC and the WNLA had forced out private contractors and gained control over recruitment.[40]

The State cooperated with the mining industry to fashion a national economy in which rural households were dependent upon remittances from migrant workers. The creation of a land-poor peasantry forced women into domestic service and farm work and men onto the mines. Low wages meant that miners rarely achieved their goal of economic independence and had to return again and again to Johannesburg. When confronted with demands for higher wages, the industry argued that any increase would reduce the labour supply because recruits would be able to meet their financial needs in shorter working periods.[41] The mining houses were so successful that, as Francis Wilson has shown, the wages of black miners did not rise between 1910 and 1970.[42]

In contrast to migrant labour in the USA, which is characterised by seasonal fluctuations, especially in agriculture, oscillating migration was persistent in its demand for workers. It was also distinctive in extent: 95 per cent of black miners were migrants.[43] The pull of the mines on labour-sending areas was accelerated by recurrent droughts and the rinderpest (cattle disease), which destroyed most of the herds in the Eastern Cape. Miners and their communities adapted to oscillating migration in subtle and effective ways, but the human costs were profound.[44] The prolonged absences of husbands and fathers damaged family life, often leading to marriage breakdown and poverty.[45] Poverty ruined the health of families, and by the late 1920s tuberculosis was well established in the Transkei and the Ciskei, areas from which the mines drew much of their workforce.[46]

On each mine one or two shafts were sunk and at regular intervals horizontal tunnels or drives would be blasted to reach the ore. During stoping[47] the ore face was drilled then blasted, and the ore lashed or shovelled manually into trucks then conveyed by tram to the shafts. Every aspect of the mining process from shaft development to tramming created dust. Although the labour system was racialised, there was little reference to race in the miners' phthisis legislation or the mines acts. Instead, those laws were framed in terms of two categories: miners (whites) and native

[40] Jeeves, *Migrant Labour in South Africa's Mining Economy.*

[41] Yette Glass, 'The African in the Mining Industry,' in *1921–1971 Mine Medical Officers' Association of South Africa: Proceedings of the Jubilee Congress of the Association, 23 to 25 March, 1971*: 10; see also Francis Wilson, *Migrant Labour in South Africa* (Johannesburg: South African Council of Churches, 1972).

[42] Black wages were probably lower in 1969 than they had been in 1911. Francis Wilson, *Labour in the South African Gold Mines 1911–1969* (Cambridge: Cambridge University Press, 1972): 46.

[43] Commission of Inquiry into Safety and Health in the Mining Industry, Braamfontein, Johannesburg, 1 August 1994, Testimony, vol. 8: 734–36.

[44] Dunbar Moodie, *Going for Gold: Men, Mines, and Migration* (Berkeley: University California Press, 1994).

[45] Trapido, 'The Burden of Occupational Lung Disease': 69.

[46] Packard, *White Plague, Black Labor*, 121.

[47] Stoping is the removal of the wanted ore from an underground mine leaving behind an open space known as a stope.

labourers (blacks), the former designated 'skilled' and the latter 'unskilled'. They did different jobs for different rates of pay and for different contract periods. Whites worked in supervisory roles, while blacks did almost all the manual labour. That work including tunnelling, preparing and drilling the holes for blasting (hammer boys), and loading and moving the skips of ore (lashers). Each of those tasks was dangerous in terms of traumatic injury. They also involved the heaviest dust exposures.

Whites and blacks worked under different compensation regimes, were subject to different forms of medical surveillance, and received different medical care, with sanatoria for white silicotics and repatriation for blacks. The nomenclature of 'miners' and 'native labourers' pervaded both official discourse and South African science and influenced the way the industry was viewed from the outside. The minutes of the 1930 Silicosis Conference in Johannesburg suggest that overseas delegates were oblivious to segregation on the mines, and the editors of the ILO's 1937 review of the South African compensation system seem to have been under the same illusion.[48]

The Transvaal Chamber of Mines was solidary and extremely powerful. The fixed price of gold meant that its members did not compete against each other for a share of a finite market. Consequently, the interests of the individual mining houses tended to coincide. In contrast to the asbestos or tobacco companies, which have faced their own health crises, South Africa's gold mines were unified in their goals and able to pursue an agreed-upon set of policies.[49] The Chamber was a politically sophisticated organisation that employed various techniques to counter its critics, including representations to government, press releases, advertisements and submissions to public inquiries. The Department of Mines shared the Chamber's interest in maintaining the industry's profitability, and there were frequent exchanges of personnel between mines management and government. An intimate relationship between the industry and the state was also fostered by the mining legislation. No miner could work underground without a certificate of fitness issued by a government approved authority.

The Chamber also had an intimate relationship with the research community. The South African Institute of Medical Research (SAIMR), founded in 1912, was an initiative of the WNLA, which had persuaded government to share the costs of research into pneumonia and miners' phthisis. The WNLA provided £40,000 for a suitable building and equipment and shared the maintenance costs,[50] and the research agenda was set

[48] *Workmen's Compensation for Silicosis in the Union of South Africa, Great Britain, and Germany,* International Labour Office, Studies and Reports, Series F (Industrial Hygiene) no.16 (London, 1937).
[49] Statement of evidence by the Transvaal Chamber of Mines in Industrial Legislation Commission of Enquiry 1948–1956, NCR 543 1 & 2: 3.
[50] Memo: The Incidents and Correspondence which led to the formation of the South African Institute of Medical Research, N. O'K. Webber, Johannesburg, 23 July 1914. South African Institute for Medical Research 17/9/1912 to 24/7/1918, WNLA 144, Teba Archives, University of Johannesburg.

by the Chamber. The research community was close-knit, and the same cast of experts appeared at departmental inquiries and the numerous commissions. Specialists often moved back and forth between private and public employ. During a career that spanned the period from 1913 until the 1960s, Dr A. J. Orenstein was variously head of sanitation for Rand Mines Ltd, chair of the Chamber's powerful Gold Producers' Committee and from 1956 the inaugural director of the Pneumoconiosis Research Unit in the Department of Mines.

By 1916 a system of medical surveillance, prevention and compensation had been created. The Miners' Phthisis Medical Bureau (later the Silicosis Medical Bureau), established by the Chamber and the State, made awards from which the official disease rates were constructed. Its system of pre-employment medicals was widely admired. At an ILO conference on silicosis in Geneva in 1938, one delegate singled out South Africa as having the most rigorous system in the world. He concluded that 'if criteria of medical examination as strict as those applied in South Africa were to be adopted in Belgium, conditions would become quite impossible, as it would immediately lead to a shortage of workers'.[51] The gold mines were subject to more stringent regulation and review than any other part of South Africa's labour system. There was little regulation of factories and virtually none of the farms or the domestic sector in which a large number of black women were employed.[52] In that context, how was it possible for widespread disease among such a large workforce to go unnoticed?

In Anglo American's annual report for 1950, the chairman told shareholders that for both humanitarian and practical reasons the company was committed to improving the health of its workforce. Anglo American had introduced mass miniature radiography and had set up clinics at mine hostels. The new hospital at Welkom was to be one of the best equipped in South Africa. 'It is well established,' the chairman wrote, 'that Natives return to their kraals generally in far better physical condition after their periods of service with the mining industry'.[53] Indeed, those benefits were supposed to have extended to migrant workers drawn from across Southern Africa. According to one industry historian A. P. Cartwright, 'It is of importance to many of the African communities south of the Equator that so many of their men return to their villages after a sojourn in South

---

[51] *Silicosis: Proceedings ILO Conference 1938,* International Labour Office Studies and Reports (London: International Labour Office, 1938): 77.

[52] As recently as 1976 the Erasmus Commission found that over 70 per cent of South Africa's eight million workers were inadequately covered by occupational health and safety legislation. *Report of the Commission of Enquiry on Occupational Health* (Pretoria: Government Printer, 1976): 95–96. In 1983 just 81 inspectors were responsible for worker safety in thirty-five thousand factories. Anthony Zwi, Sharon Fonn and Malcolm Steinberg, 'Occupational Health and Safety in South Africa: The Perspectives of Capital, State, and Unions,' *Social Science and Medicine* 27 (1988): 691–702.

[53] 'Chairman's Report,' in *Annual Report of Anglo American for the Year Ended 31 December 1950*: 3–4.

Africa in far better physical condition than most of them were when they left home.'[54]

For almost a hundred years South Africa's gold mines claimed to be leading the world in safety, medical surveillance and compensation, but a careful reading of the history suggests that was an illusion. The industry's failure to create safe workplaces and to compensate migrant workers for occupational disease underpinned its commercial success and allowed the costs of production to be shifted to rural communities.

[54] A. P. Cartwright, *Doctors of the Mines: A History of the Work of Mine Medical Officers* (Cape Town: Purnell and Sons, 1971): 3.

2
Creating a Medical System

In May 1901 the gold mines, which had been closed during the South African War, reopened. Many of the Cornish miners who had left for Britain at the beginning of the war had never returned and, according to the government mining engineer for the Transvaal, in the interim more than two hundred former rock drillers had died of silicosis. The report created adverse publicity in the British press and questions were asked in the House of Commons. In February 1902 a British parliamentary delegation visited South Africa to investigate the incidence of silicosis and one MP, Gilbert Parker, wrote to the Chamber of Mines, 'I am convinced that the importation of natives from the Zambezi valley and from Central Africa ought to be stopped, the percentage of deaths amongst the natives on mines [from pneumonia] being seriously high chiefly because of the mortality amongst these particular natives.' In response, the commissioner for native affairs, Godfrey Lagden, met with the Chamber to discuss what was clearly a health crisis. A committee was appointed that included Drs L. G. Irvine, D. MacAuley and Andrew Watt.

The committee report identified a serious problem with the mortality rate at 54.5 per 1,000. It noted: 'Native workers are particularly susceptible to this disease [pneumonia] and the conditions of mining work favour its incidence.'[1] Pneumonia was caused by a combination of fatigue, crowded compounds, poor nutrition, and inadequate clothing. Exposure to cold and damp were of particular concern as was scurvy with one mine medical officer's report showing its presence in over 10 per cent of fatal cases.[2] In contrast to silicosis, for which most deaths occurred among repatriated black miners in remote rural areas, pneumonia killed recruits in the compounds under the gaze of the Department of Mines. The superior

[1] Cartwright *Doctors of the Mines*: 16, 17–18.
[2] Mine rations were deficient in protein and vegetables. In 1906 the Department of Native Affairs gazetted a standard diet that included meat, vegetables and mealie meal (maize). Further improvements during the 1920s saw a decline in deaths from malnutrition. See Simons, 'Migratory Labour': 8.

general health and working and living conditions enjoyed by white miners meant that the victims of pneumonia were always black.

The medical crisis had an impact on recruitment, and the mines struggled to attract labour. By August 1902 the Rand mines were employing only thirty-seven thousand, well below the hundred thousand men required for full production.[3] As a result, on 7 November 1902 the high commissioner for South Africa, Lord Milner, appointed a commission (which became known as the Weldon Commission) to investigate the causes and extent of silicosis.[4]

The Weldon Commission's report, submitted in 1903, found that silicosis was a major problem among all classes of mine labour but especially amongst rock drillers, who had an average working life of just seven years. (In fact, it was probably closer to four, as the dust levels in Johannesburg mines were particularly hazardous.)[5] It correctly identified the cause as the fine silica dust generated by pneumatic drills and the use of gelignite for blasting.[6] The remedies it recommended were more stringent blasting regulations, wet drilling, mechanical ventilation and water sprays for laying dust. Weldon made barely a reference to black miners or the toll that pneumonia was taking on migrant labour. Its report led to only a small reduction in dust levels, and the British scientists J. S. Haldane and Sir Thomas Oliver continued to pressure the British government to improve conditions on the Rand.

The mortality rate from pneumonia, particularly among men recruited from the tropical North (the region north of 22 degrees South latitude), remained very high. On the Simmer and Jack mine in 1904 it stood at 360 per 1,000.[7] One obvious remedy was to improve conditions in the compounds but the mining houses resisted change and despite prolonged discussions between industry and the Milner administration the deaths continued.[8] Before 1910 the silicosis burden was probably spread evenly throughout the workforce; Weldon's finding that all classes of labour were at risk was endorsed by the Miners' Phthisis Commission of 1912.[9] After 1912 a reduction in dust levels achieved through blasting regulations, the use of water for dampening down and the introduction of mechanical ventilation led to a lowering of the silicosis threat. According to the official data in the period from 1916 to 1935 there was a constant fall in the

[3] Cartwright *Doctors of the Mines*: 7.

[4] Within six days the British Home Office in London announced the establishment of the Haldane Commission to study silicosis among Cornish miners. That decision was overdue, as the death rate among tin miners in the UK had long been a matter of public concern. Katz, *The White Death*: 25.

[5] Katz, *The White Death*: 4.

[6] *Report of the Miners' Phthisis Commission 1902–1903* (Pretoria: Government Printer, 1903): xxi–xxii.

[7] Cartwright *Doctors of the Mines*: 19.

[8] Cartwright *Doctors of the Mines*: 20.

[9] *Report of a Commission of Inquiry into Miners' Phthisis and Pulmonary Tuberculosis* (Cape Town: Government Printer, 1912): 6–8.

numbers.[10] While the accuracy of the data is questionable, we can be certain that silicosis was transformed from an acute to a chronic disease. From that point the divide in the health profiles of white and black miners became more pronounced. Few white silicotics contracted tuberculosis, and those with simple silicosis were given surface jobs, pensions and retraining. The small number who developed tuberculosis went to Springkell Sanatorium, established by the Chamber and the government for their care. In contrast, few blacks were diagnosed with simple silicosis, and the far larger number who developed tuberculosis or silicosis with tuberculosis were repatriated, many without compensation.[11] As the threat of acute silicosis receded, the issue of tuberculosis became more prominent.

The crises of acute silicosis, pneumonia and chronic silicosis overlapped in time. Each was triggered by dust in the mines, conditions in the compounds and the industry's insatiable demand for labour. Eventually, under pressure from the imperial authorities in 1913, recruitment of labour from the tropical North was banned, saving thousands of lives by removing the most vulnerable men from risk work. The Chamber also experimented with vaccines developed by Spencer Lister and the British bacteriologist Sir Almroth Wright.[12]

The legislative framework

Medical surveillance and compensation in the South African gold mines began with the Miners' Phthisis Act of 1911. Compulsory medical examinations were to ensure the fitness of recruits for gruelling labour underground and prevent the spread of tuberculosis, thereby reducing the cost of compensating men who came to the mines with weak lungs. The sophistication of the system increased with subsequent acts. In terms of scale, longevity and the degree to which it was racialised, the South African medical system was unique.

The Mines and Miners' Phthisis Acts from 1911 to 1925 established a system of workplace regulation, medical surveillance and compensation that in principle subjected the industry to external review. The legislation had a number of notable features. South Africa was the first state to compensate for silicosis and tuberculosis as occupational diseases, and it was the first to introduce medical certification for hard rock miners. The

---

[10] *The Prevention of Silicosis*: 268–74 and 25.

[11] *Report of the Miners' Phthisis Acts Commission, 1941–1943* (Pretoria: Government Printer, 1943); Letter by Dr L. Bostock, District Manager of the Witwatersrand Native Labour Association to the General Manager, Transvaal Chamber of Mines, Johannesburg 20th November 1924, quoted in full in *Proceedings of the Transvaal Mine Medical Officers' Association* 4, no. 11 (1925): 5.

[12] This initiative was used by the WNLA and the NRC in their negotiations with colonial administrations. See Marais Malan, *In Quest of Health: The South African Institute of Medical Research, 1912–1973* (Johannesburg: Lowry Publishers, 1988): 95–111.

Miners' Phthisis Medical Bureau lay at the system's heart. It was responsible for all pre-employment, periodic, exit and compensation examinations of white miners and for the medical examinations of black miners conducted at individual mines or at the WNLA compound.

Miners' Phthisis Act No. 44 of 1916 made tuberculosis compensatible and barred any person with tuberculosis from underground work. It defined miners' phthisis as silicosis, and tuberculosis was defined as tuberculosis of the lungs or respiratory organs. No diagnostic criteria were given. The Act made the Miners' Phthisis Medical Bureau responsible for ensuring that medical examinations were performed by doctors approved and gazetted by the minister and for compiling data on the incidence of disease. Although it appeared to bring medical reviews under the authority of the Bureau and therefore the State, the act decentralised the system. White miners were examined at the Bureau; black miners were examined by mine medical officers or at the WNLA compound. William Gemmill, who was a dominant figure in the Chamber, explained how the system worked: 'Every Mine Medical officer is by law an officer of the Miners' Phthisis Medical Bureau, and is subject to the instructions of the Bureau in all Miners' Phthisis matters affecting natives employed by the Mines.'[13] In theory he was correct, but as with all such legislation there was a gulf between what was prescribed and what actually happened in the workplace.[14] In practice, medical examinations of black miners were controlled by employers. This structure was retained in the acts of 1919 and 1925 and became the hallmark of mine medicine and compensation.

It was easy for government to justify outsourcing medical examinations to employers. By 1915 there were over one hundred thousand black miners on the Rand, all of whom were subject to compulsory examination. There was also a high labour turnover, which increased the workload. The cost for the state in running that system would have been prohibitive and would no doubt have been a source of conflict between government, the mines and the white electorate. For the mining industry there was an even more important consideration. Within months of the passage of the Act of 1912 compensation costs began rising, and the industry was keen to minimise the number of claims. The most important question for employers was in regard to who controlled the medical review system and therefore the referrals for compensation. The industry was willing to bear the cost of medical examinations so long as it also controlled compensation referrals to the Miners' Phthisis Medical Bureau.

Miners' Phthisis Act No. 40 of 1919 formalised the system. The minister, in consultation with the board, had the authority to issue regulations regarding the duties of medical practitioners and the conduct of examinations. When a miner who had worked underground for a period

---

[13] William Gemmill, General Manager Tropical Areas, Note for Mr K. Lambert Hall, Secretary, Nyasaland, Northern and Southern Rhodesia Inter-territorial Conference, Salisbury, 20 December 1940, Emigrant Labour Governors Survey M2/3/19, Malawi National Archives.
[14] Simons, 'Migratory Labour': 27–43.

exceeding one month left employment, he was to be given a final examination by the Bureau or by a designated medical practitioner. The act's
most important innovation was to lay down the criteria for diagnosing
tuberculosis. A miner was to be compensated only where the tubercle
bacillus was present in his sputum or he was suffering from a serious
impairment of working capacity. There was no pathology testing at the
WNLA in 1919 and, indeed, there were no such facilities at the WNLA
hospital as late as the early 1950s.[15] Nor did the act offer guidelines as to
what constituted serious impairment.

A medical practitioner who found symptoms of tuberculosis or silicosis in a black miner was required within three days to report the case to
the Bureau. If in conducting a post-mortem examination a medical officer
found evidence of silicosis, he was required to send a report and the lungs
to the Bureau. Every member of the Bureau and every gazetted medical
practitioner had the right of entry to scheduled mines (mines deemed
hazardous). Finally, employers who failed to carry out any order given by
competent authority for the prevention of silicosis were liable on conviction to a maximum fine of £500 or imprisonment for up to two years.

Miners' Phthisis Acts Consolidation Act No. 35 of 1925 tightened the
compensation provisions to the disadvantage of black miners. Benefits for
tuberculosis without silicosis were extended to former miners so long as
they were diagnosed within six months of having ceased work. Tuberculosis was added to the provision that an employer who failed to carry out
any order for disease prevention would be fined. The Act is impressive in
terms of its ambit and in the severity of its penal provisions. Anyone reading
it would assume that the mines were subject to rigorous state regulation.

The tenor of the legislation becomes clearer, however, when viewed in
light of the state's response to pneumonia. While pneumonia was probably the outstanding cause of death among black recruits, it was never
made the subject of a commission of inquiry or declared an occupational
disease. A sustained decline in the mortality rate from pneumonia did
occur after the ban on the recruitment of Tropical labour was imposed in
1913. Even so, between 1933 and 1938 pneumonia accounted for between
29.65 and 37.87 per cent of deaths on the mines from disease.[16] Such a
lack of legislative reform suggests that South African policy makers were
preoccupied with silicosis because it was a threat to white miners.

Mine medical officers

The medical system of South Africa's gold mines had no parallel in the
colonial world. The United Fruit Company in Central America, the Union
Minière in Katanga (the Congo) and the tea and sugar plantations in India,
Mauritius and Sumatra were similar in scale, but the risks faced by plan-

---

[15] Interview with Dr Gerrit Schepers, Grand Rapids, Virginia, 23–28 October 2010.
[16] *Proceedings of the Transvaal Mine Medical Officers' Association* 19, no. 214 (1940): 229–30.

tation workers were in no sense comparable to those on the Rand.[17]
Nowhere was such a large group of migrant workers concentrated in a
compact area in which medical officers met regularly and a medical
service fell under the aegis of a single employers' association.

Mine medical officers had to deal with a wide range of illnesses,
including silicosis, tuberculosis, pneumonia, enteric fever, malaria and
meningitis, and with parasitic infestations such as hookworm. Traumatic
injuries were common, and in addition to the five hundred or so annual
accidental deaths large numbers of miners required emergency surgery.
Every doctor had a heavy administrative load requiring an understanding
of the miners' phthisis legislation and the compensation system. The daily
round included reading hundreds of X-rays, routine laboratory work, and
conducting pre-employment and exit medical examinations and post-
mortems. They were also responsible for sanitation in the compounds.

The system of surveillance began with pre-employment medical examinations, and from August 1914 periodic examinations of black miners
were introduced. With the passage of the Miners' Phthisis Act of 1916
examinations were performed every three months. It was a massive undertaking that Dr A. J. Orenstein, then the head of sanitation for Rand Mines
Ltd., characterised as 'medical mass production'.[18] Over time the system
became more refined, and the Act of 1925 made the employment of full-
time medical officers compulsory on the larger mines. The pay was low,
and medical officers usually combined their duties with an extensive
private practice. Mine medicine was a difficult career made no easier by
the tension between the industry's desire for labour and its determination
to control costs.[19]

The medical service consisted of a small number of senior officers
supported by a group of young graduates who were using the mines as a
stepping stone to a better career. It tended to be a corps of transients, and
in the period between 1929 and 1935, for example, thirteen out of thirty-
four doctors resigned from the service.[20] A full-time medical officer on a
medium-sized mine would be responsible for around 6,000 men.[21] The
workload of the five full-time physicians at the WNLA hospital was also
crushing. In addition to being responsible for more than two hundred and
fifty hospital patients, each doctor examined between three hundred and
one thousand two hundred black miners a day.[22] The corps' status was

---

[17] F. Daubenton, 'Training and Specialisation of Mine Medical Officers' *Proceedings of the Transvaal Mine Medical Officers' Association* 15, no. 164 (1935): 75–76.

[18] *Proceedings of the Transvaal Mine Medical Officers' Association* 15, no. 164 (1935): 80.

[19] For an overview of the system see Beris Penrose, 'Medical Monitoring and Silicosis in Metal Miners: 1910–1940', *Labour History Review* 69, no.3 (2004): 285–303.

[20] Daubenton, 'Training and Specialisation of Mine Medical Officers' *Proceedings of the Transvaal Mine Medical Officers' Association* 15, no. 164 (1935): 76.

[21] On occasion it could be as high as twelve thousand. See Dr Williams in discussion, *Proceedings of the Transvaal Mine Medical Officers' Association* 15, no. 164 (1935): 81.

[22] 'Reservations by Mr. W. Boshoff' in *Report of the Miners' Phthisis Commission of Enquiry, Part Two* (Pretoria: Government Printer, 1930), 100.

such that mine medical officers were referred to by their Johannesburg colleagues as 'Kaffir Doctors'.[23]

Miners recruited within South Africa were given a cursory examination at the point of recruitment and then examined at the Native Recruiting Corporation (NRC) compound at Driehoek. Before going underground they were examined again by a mine medical officer. East Coasters (migrants from Mozambique) were assembled by the WNLA at Ressano Garcia and examined by a Portuguese doctor before being transported to Johannesburg. Twice-weekly gangs of five to six hundred men would arrive at the WNLA compound, and those judged fit were sent on to the mines. The rejection rates of East Coasters at Ressano Garcia and Johannesburg were around 1 per cent.[24] It was costly to transport labour from remote rural areas, and therefore it is not surprising that the rejection rates were so low. In addition to the men recruited by the NRC and the WNLA, a large number of so-called voluntary workers, some of whom had already been rejected by the WNLA on medical grounds, presented themselves at individual mines. The chair of the Miners' Phthisis Prevention Committee, R. N. Kotze, warned the Chamber of Mines in 1914 that some mines were so desperate for labour that not all voluntary recruits were being examined.[25]

The size of the workforce meant that mine doctors had to perform each examination in less than two minutes.[26] Workers were lined up naked and weighed and then had their chests examined with a stethoscope. Doubtful cases were set aside for a more thorough review.[27] In January 1916 Dr George Albert Turner, the senior medical officer for the WNLA, told a parliamentary select committee that he was confident in his ability to examine just under a thousand men in a three-hour period: 'I have been at this work for ten years, and I think, provided the native is stripped stark naked, I can size him up pretty well as to what he is worth.'[28] Prior to the introduction of mass miniature X-rays after World War II, the most important tool for detecting lung disease was periodic weighing. In 1916 the

[23] Dr Miller, 'Presidents Address' *Proceedings of the Transvaal Mine Medical Officers' Association* 18, no. 206 (1939): 147.

[24] During 1920, of the 60,164 East Coasters recruited 1.1 per cent were rejected at Ressano Garcia and 0.76 per cent were rejected at Johannesburg. See A. I. Girdwood, 'Tuberculosis: Examination of East Coast Recruits' *Proceedings of the Transvaal Mine Medical Officers' Association* 1, no. 5 (1921): 3. See also George Albert Turner, evidence in Miners' Phthisis Working of Acts Parliamentary Select Committee 1916, SC 10–15 Third Report AN 4923 (Cape Town: Parliamentary Library, 1915): 592.

[25] Letter from R. N. Kotze, Chairman Miners' Phthisis Prevention Committee, to John Munro, President of the Transvaal Chamber of Mines, Johannesburg, 26 February 1914. WNLA 173 Tuberculosis November 1913 to January 1919, Teba Archives, University of Johannesburg.

[26] W. Watkins-Pitchford, 'The Silicosis of the South African Gold Mines, and the Changes Produced in it by Legislative and Administrative Efforts' *Journal of Industrial Hygiene* 9, no. 4 (1927): 129.

[27] 'Discussion of Tuberculosis' *Proceedings of the Transvaal Mine Medical Officers' Association* 5, no. 7 (1926): 5.

[28] Turner, evidence in Miners' Phthisis Working of Acts Parliamentary Select Committee 1916, 590–91.

Miners' Phthisis Medical Bureau made weighing compulsory, and initially the mines were required to weigh workers every three months. That was soon changed to every thirty days. Under the guidelines men who had lost five pounds or more between two weighings or six pounds over three consecutive weighings received a more detailed examination,[29] which in a minority of cases included an X-ray.

Each new Miners' Phthisis Act increased the number of examinations and thereby the workload. In 1924 there were only thirty-seven medical officers on the Rand, of whom thirteen were full-time. A minimum of seven hundred thousand examinations were performed annually, giving each physician a caseload of nineteen thousand. The workloads at the Miners' Phthisis Bureau were less arduous, but even so each intern would conduct around four thousand two hundred examinations of white miners a year.[30] The lack of doctors in the rural areas and their inadequate training meant that some recruits with serious illness were forwarded to Johannesburg. Dr J. F. Young remarked at a Mine Medical Officers Association meeting in June 1925 that there were thirty medical officers in the Transkei but none had ever worked on a mine. Such doctors rarely stayed in the job.[31]

One of the problems facing doctors was the lack of a standard of fitness. The standards used in assessing recruits depended on the production needs of the mines, not on the workers' health.[32] Those decisions were left in the hands of individual doctors, who were asked to keep disease out of the mines and maintain a full complement of labour. As Young reminded his colleagues at one point, 'With every mine shrieking out for labour what are we to do? The boys must be developed and the only place to develop them is where they can get decent food, on the mines. They will not develop in the Transkei where they are half starved.'[33] During discussion with his fellow medical officers in 1925 Dr W. Skaife admitted that the mines were so desperate for labour they would accept recruits just to keep going. Many recruits, he said, arrived in a state of malnutrition, and 'it would pay to fatten up natives in poor general condition before they are sent to the mines'.[34]

[29] 'Miners' Phthisis Act No. 35 (1925) Condensed Précis of Regulations and Procedure (Natives and Non-Europeans) for Mine Medical Officers' *Proceedings of the Transvaal Mine Medical Officers' Association* 6 no. 4 (1924): 4. According to Dr Oluf Martiny, who joined the WNLA in 1954 and eventually became medical director, the same methods were being used in the 1970s. He recalls one particularly busy morning when he and his seven colleagues examined twelve thousand recruits. Interview with Dr Oluf Martiny, Forest Town, Johannesburg, 27 April 2011.

[30] *Report of the Miners' Phthisis Medical Bureau for the Twelve Months ending July 31, 1924* (Pretoria: Government Printer, 1925): 34.

[31] *Proceedings of the Transvaal Mine Medical Officers' Association* 5, no. 2 (1925): 6.

[32] 'Discussion of Tuberculosis'.

[33] *Proceedings of the Transvaal Mine Medical Officers' Association* 1, no. 5 (1921): 10.

[34] Dr W. Skaife, 'Detection and Prevention of Tuberculosis' *Proceedings of the Transvaal Mine Medical Officers' Association* vol. V, No. 5 September–October 1925: 1–4.

Shortcomings of the surveillance system

The medical examinations of workers were a massive undertaking, and it is not surprising that within months of the passage of the first Miners' Phthisis Act in 1911 their usefulness was being questioned. The mines wanted labour, but management was unwilling to make the necessary investment to exclude recruits who, although ill, could pass a cursory examination. The Parliamentary Select Committee on the Miners' Phthisis Acts of 1916 was highly critical of the procedures, referring to them as 'inspections' or 'parades'.[35]

The South African research community agreed with their US, UK and Australian counterparts that it was difficult to diagnose lung disease in hard rock miners. The 1912 Commission of Inquiry into Miners' Phthisis noted problems in diagnosing tuberculosis in a fibroid lung, and that an accurate diagnosis required the patient's work history and a careful clinical examination in conjunction with a chest X-ray.[36] Those complexities were compounded by the legal criteria laid down for diagnosis and by the incompetence of some medical officers. The early data show dramatic variations in the incidence of tuberculosis on individual mines. In 1916, for example, the Crown Mines lodged 110 claims for tuberculosis while the East Rand Proprietary Mines, which had a comparable number of workers, made only eight claims. Eleven of the Crown group diagnosed with tuberculosis had worked one month or less, and in 56 cases the recruits had worked less than four months. Those figures suggest that the initial examinations at Crown were inadequate.[37]

Weight loss was the first method of diagnosis used throughout the industry and the one upon which mine doctors most depended. That most of the deaths at the WNLA compound were from tuberculosis suggested to A. I. Girdwood that monthly weighing was unreliable.[38] In September 1925 Dr A. Smith pointed out at a mine medical officers meeting that in addition to tuberculosis there were various causes of weight loss. Because of the intense work required, the weight of a hammer boy, for example, could fluctuate by four or five pounds in a few hours.[39] Furthermore, there was no reliable standard for what constituted a short weight. Dr Donaldson commented that a weight standard would penalise the industry; on one

[35] *Third and Final Report of the Select Committee on Working of Miners' Phthisis Acts,* Union of South Africa Select Committee 10–15, AN 492 April 1916, National Archives of South Africa: xxvi.

[36] *Report of a Commission of Inquiry into Miners' Phthisis and Pulmonary Tuberculosis*: 11.

[37] Memorandum: Miners' Phthisis & Tuberculosis in Natives from Legal Advisor 27th February, 1917. WNLA 144/2 Miners' Phthisis Act September 1916 to November 1923, Teba Archives, University of Johannesburg, Teba Archives: 2–4.

[38] *Proceedings of the Transvaal Mine Medical Officers' Association* 1, no. 7 (1921): 3.

[39] A. Smith, 'Weighing' *Proceedings of the Transvaal Mine Medical Officers' Association* 5, no. 5 (1925): 5.

mine where he had enforced the removal of men who had suddenly lost five pounds the mine was temporarily and, in his view, unnecessarily deprived of five hundred workers.[40]

The mine doctors had founded an association known as the Transvaal Mine Medical Officers' Association (MMOA) in 1921. Its first elected officers were Dr H. T. H. Butt of the Johannesburg Consolidated Investment Company, president, Dr A. J. Orenstein of Rand Mines, vice-president, and Dr A. I. Girdwood, the chief medical officer of WNLA, secretary and treasurer. The association received financial support from the Chamber, while the WNLA and the NRC funded the publication of its monthly proceedings.[41] In 1925 the association sent a questionnaire to its members to gauge the diagnostic scheme's effectiveness. The responses showed that in 47 per cent of cases tuberculosis had been discovered through weight loss and in 15 per cent it had been discovered by accident. In the remainder of the cases tuberculosis had been identified following 'acute disease', meaning that by the time of diagnosis the miner was dying.[42] Subsequent research suggested that although weight loss was usually present in some patients with advanced disease that was not always the case.[43]

Dr W. Watkins-Pitchford, then director of the South African Institute of Medical Research (SAIMR), was unconvinced of the usefulness of weight loss as a diagnostic tool. In 1927 he wrote: 'Experience has shown … that, for one reason or another, this system does not lead to the detection of more than about 50% of the cases of simple tuberculosis and of tuberculosis with silicosis which are finally discovered.' Consequently, many cases of tuberculosis were not detected until the miners had become too ill to work. Watkins-Pitchford concluded, 'In order to mitigate the mischief of which the overlooked native "carrier" of tuberculous infection is the source, I have lately proposed utilising large-scale X-ray examination by means of the fluorescent screen.'[44]

The other factor that compromised weighing as a diagnostic tool was the condition of recruits on arrival. A study of over twenty thousand migrant workers at the beginning and end of their contracts found that the average weight was 132.4 pounds on entry and 135.7 pounds on discharge.[45] Twenty years later a parliamentary committee commented on the almost universal weight gain among migrant workers during the first months of employment. Malnutrition was rife in the reserves, and recruits

---

[40] *Proceedings of the Transvaal Mine Medical Officers' Association* 5, no. 5 (1925): 8.

[41] *Proceedings of the Transvaal Mine Medical Officers' Association* 1, no. 12 (1922): 4. In 2002 the MMOA changed its name to the Mining Medical and Other Health Care Professionals' Association, and its secretariat is still located in the Chamber's headquarters in Johannesburg.

[42] *Proceedings of the Transvaal Mine Medical Officers' Association* 5, no. 5 (1925): 7.

[43] N. R. A. MacColl, 'The Early Diagnosis of Tuberculosis,' *Proceedings of the Transvaal Mine Medical Officers' Association* 20, no. 218 (1940): 1–2.

[44] Watkins-Pitchford, 'The Silicosis of the South African Gold Mines': 128, 129.

[45] *Tuberculosis in South African Natives with Special Reference to the Disease Amongst the Mine Labourers on the Witwatersrand* (Johannesburg: South African Institute for Medical Research, 1932): 77.

often gained weight because their diet had suddenly improved.[46]

Stethoscopic examinations were also of limited value. Many active tuberculosis cases initially showed little stethoscopic evidence of disease. The MMOA often discussed diagnostic techniques. At a meeting in July 1925 Dr Stoney commented on the difficulties he was having in conducting examinations. At the mine where he worked a train carrying ore to the mill passed close to the hospital and made such a horrible row that he found it almost impossible to listen to the miners' breathing.[47] In March 1943 Dr L. F. Dangerfield noted that making a minimum of two hundred stethoscopic examinations a day with the object of discovering early tuberculosis was wasted effort.[48]

Under the Act of 1925 compensation for tuberculosis was based on the presence of tubercle bacillus in the sputum or marked physical incapacity. Because positive sputum was the exception rather than the rule in early-stage infection, those prescriptions added to the problems of diagnosis. It was common for a miner to have tubercle bacilli in his sputum and physical signs suggestive of tuberculosis but return a negative X-ray plate. Conversely, a miner might display radiological evidence of disease without appreciable physical symptoms. Dr F. Retief spoke at length on the challenges of diagnosis at a MMOA meeting in February 1940. He noted that a single sputum sample in a suspected tuberculosis case was often inconclusive and that repeated examinations should be made. Where physical and radiological tests were suggestive of infection, it might be necessary to do as many as six or seven examinations before the bacilli were found.[49] Medical officers who carried out thousands of examinations each week were incapable of following such procedures.

The South African gold mines were the first industry to routinely X-ray employees. The report of the 1912 Commission featured a cohort of miners from the Simmer and Jack Mine Hospital who were X-rayed by Dr A. Watt of Rand Mutual Assurance Company.[50] It was the first study of working miners of its kind and a landmark in South African medical history. In 1916 the Miners' Phthisis Prevention Committee commented that while an X-ray provided the most reliable single piece of evidence an accurate diagnosis required both radiographic appearance and a clinical

---

[46] *Report of the Departmental Committee of Enquiry into the Relationship Between Silicosis and Pulmonary Disability and the Relationship Between Pneumoconiosis and Tuberculosis*, Part 2, *The Relationship Between Pneumoconiosis and Tuberculosis*. Departmental Committee to Inquire into Definition of Silicosis & Chest Diseases, Departmental Committee, 1954, South African National Archives F 33\671 Treasury: 36.

[47] *Proceedings of the Transvaal Mine Medical Officers' Association* 5, no.3 (1925): 3.

[48] L. F. Dangerfield, 'Pulmonary Tuberculosis in South Africa and the Problem of the Native Mine Labourer' *Proceedings of the Transvaal Mine Medical Officers' Association* 22, no. 249 (1943): 173.

[49] F. Retief, 'The "Clinical Side" of Tuberculosis' *Proceedings of the Transvaal Mine Medical Officers' Association* 19, no. 215 (1940): 237.

[50] *Report of a Commission of Inquiry into Miners' Phthisis and Pulmonary Tuberculosis*: 12–26.

examination.[51] After 1916, however, the Miners' Phthisis Medical Bureau relied almost entirely upon X-rays in assessing white miners. The use of X-rays in the examination of black miners at entry and periodic examinations was discussed by the MMOA on a number of occasions. The Act of 1919 extended X-rays of white miners to periodic examinations, but the suggestion that the same system be used with blacks was rejected because of cost.[52] At the beginning of 1930 the WNLA made an attempt at providing X-rays for black miners, but the scheme soon ran into problems; the number of cases to review made the workload overwhelming. At the request of the MMOA, the Chamber ended the trial prematurely.[53]

Much to the annoyance of the Rand Mutual Assurance Company, which held policies for the major mining houses, every year a sizeable number of recruits would break down after brief periods underground. For example, a migrant worker named Sikambele, who had been repatriated following an accident at the Modderfontein Deep in June 1911, was subsequently recruited by the WNLA and returned to Johannesburg in December 1912. He worked only one shift before being diagnosed with 'marked phthisis'. When he died on 16 March 1913, a claim for £30 was lodged by the Department of Native Affairs on behalf of his estate.[54] Sikambele's was one of several cases in which claims were made by men who had worked underground for a month or less.[55] The managing secretary of Rand Mutual wrote in protest to the WNLA, pointing out that Sikambele had been under medical treatment at the WNLA's compound after his accident in 1911 and his illness should have been diagnosed at that time. The company wanted stricter pre-employment medical examinations. In December 1912 the Chamber's legal adviser, G. E. Barry, wrote an assessment of the industry's growing legal obligations: the cost of compensating black miners was rising fast, and there was a need for more stringent entry medicals to prevent the employment of sick men.[56]

In December 1920 officers from the Rand Mutual met with the Chamber to discuss medical examinations. Mr Munro of Rand Mutual pointed to

[51] *General Report of the Miners' Phthisis Prevention Committee Johannesburg 15th March 1916* (Pretoria: Government Printing & Stationery Office, 1916): 13.

[52] Packard *White Plague, Black Labor*, 183

[53] Letter from A. Percival Watkins, Transvaal Mine Medical Officers' Association, to General Manager, Transvaal Chamber of Mines, Johannesburg, 28 November 1930. WNLA 20L Diseases and Epidemics, Tuberculosis February 1923 to December 1930, Teba Archives, University of Johannesburg.

[54] Letter from W. Tudhope, Secretary Modderfontein Deep Levels Limited, to the Managing Secretary, Rand Mutual Assurance Company, Johannesburg, 12 March 1913. WNLA 138 Phthisis: Allotment of Natives January 1913 to May 1915, Teba Archives.

[55] Letter from Managing Secretary, Rand Mutual Assurance Company, Johannesburg, to The Secretary, WNLA, 20 March 1913. WNLA 138 Phthisis: Allotment of Natives January 1913 to May 1915, Teba Archives, University of Johannesburg.

[56] Memo from Mr G. E. Barry, Legal Advisor to the Chamber of Mines Compensation for Miners' Phthisis in Native Labourers under Act 19 of 1912, 18 December 1912. WNLA 138/2 Miners' Phthisis Amendment Act June 1912 to February 1916, Teba Archives, University of Johannesburg.

'cases where the boys have not worked at all after going there [to the mines]. They have been sent to the mine compound but have not been put to work in many cases, and they waste away and die in the place.' If there were stricter pre-employment medical examinations, he argued, the cost of compensation would fall as would the number of recruits.[57] While the parties agreed that the supply of labour was at stake, they seem also to have agreed that the problem was intractable.

Repatriation

One of the most fraught aspects of the medical system was the repatriation of injured or sick miners. Each repatriation required a medical officer to conduct an examination and fill out a form. If found fit to travel, the man was certified, a report was forwarded to the native protector of his home district and he was sent to the WNLA hospital.[58] By 1930 around 120 black miners awaiting repatriation were processed each week.[59] Every Tuesday morning two groups were dispatched from Booysens Railway Station, one to Ressano Garcia in Mozambique and the other to the Cape Province. A medical officer saw the trains off, but there were no doctors on board. Each train had special coaches for the bedridden; the rest travelled in third-class carriages. During the journey the men were fed and cared for by a white attendant. If a patient was unable to walk at the railhead, transport was arranged. Should a miner die en route, compensation was paid to his dependents once they had been identified. Unless a man recovered and returned to Johannesburg, repatriated miners simply disappeared.[60]

Repatriation had the full support of the Johannesburg medical estab-lishment. Writing in 1916, the Miners' Phthisis Prevention Committee concluded that repatriation was the best way to hasten the recovery of sick workers. It was also confident that the mines were not spreading tuber-culosis to 'the kraals'.[61] Not all medical officers shared the committee's view, and even before the introduction of the first phthisis legislation some raised their concerns about repatriation. Dr Alexander Frew, for example, observed in 1910 that 'from the point of view of the future labour supply of the Rand, there seems no doubt a mistake is being made in sending these invalids back to their kraals to infect future mine labour.'[62] Frew's

---

[57] Notes of Meeting between the Transvaal Chamber of Mines, Rand Mutual Assurance Company, the NRC and WNLA, Johannesburg, 20 December 1920. WNLA 138/2 Miners' Phthisis Amendment Act June 1912 to February 1916. Teba Archives: 2–5.

[58] G. A. Turner, 'Report on Tuberculosis Among Natives' undated (October 1913?), WNLA 173 Tuberculosis November 1913 to January 1916, Teba Archives, University of Johannesburg, 8.

[59] *Tuberculosis in South African Natives*: 86.

[60] Dr. Hertslet, discussion of 'The Disposal of Tuberculosis Cases' *Proceedings of the Transvaal Mine Medical Officers' Association* 2, no. 6 (1922): 1.

[61] *General Report of the Miners' Phthisis Prevention Committee*: 17–18.

[62] Alexander Frew, 'Tuberculosis among Mine Natives' *Transvaal Medical Journal* 6, no. 4 (1910): 63–64.

suspicions were supported by evidence from the Eastern Cape. In 1918 Dr Welsh of the NRC's Umtata office reported that at least 15 per cent of local men given a cursory examination were found to be infected, and he was certain that a fuller examination would bring that figure up to 20 per cent. 'After over 25 years working amongst the natives,' he wrote, 'I am quite satisfied that tuberculosis is increasing amongst them, and is likely to continue to increase.'[63] In 1926 an NRC survey of 110 men repatriated to the Transkei found that 14 were dead within a month of arriving home, 25 within three months, and a further 25 within six months, while an additional 19 were dying at the time of the survey.[64] A study from the Robinson Deep Mine showed that of the 348 tubercular miners repatriated to the Transkei between 1926 and 1928 more than 40 per cent had died in the first year. Although both studies drew on small samples, they did suggest that there was a serious problem.[65]

The MMOA spent a great deal of time discussing tuberculosis.[66] In political terms, there was possibly no more important question than where black miners acquired that disease. If infection was being brought into the mines by migrant workers, the best way of reducing disease was to improve the quality of the entry medicals. If tuberculosis was being contracted on the mines, then the problem was the synergy with silicosis, and the solution was to eliminate the dust and improve conditions on the compounds. At a MMOA meeting in August 1921 devoted to 'the disposal' of infected miners, Dr Girdwood gave a paper on tuberculosis among recruits from Mozambique north of 22 degrees South latitude. He was confident that few cases of tuberculosis or silicosis escaped detection in the initial examinations at Ressano Garcia and Johannesburg. He was also sure that pulmonary tuberculosis was uncommon among East Coasters: out of the 174,402 blacks employed during 1920, only 28 were compensated for tuberculosis or silicosis during the first two months of their contracts.[67] Dr Allen agreed with him that tuberculosis was extremely rare among new recruits: 'I think that by far the greatest majority of the affected boys give a history of previous mine work. Therefore I assume that the infection amongst the natives with Tuberculosis – I am confining myself to East Coast natives only – is acquired on the mines.'[68] During a lively discussion Allen went on to argue: 'We must not forget that the mine hospital is not a sanatorium, and for the boy's own sake I consider that he

[63] Letter from Native Recruiting Corporation, Umtata, to the Secretary Native Recruiting Corporation, Johannesburg Incidence of Tuberculosis amongst Natives, 23 August 1918, NRC 135 Miners Phthisis and Tuberculosis among Natives, Teba Archives, University of Johannesburg.

[64] *Proceedings of the Transvaal Mine Medical Officers' Association* 5, no. 7 (1926): 4.

[65] L. F. Dangerfield, 'Pulmonary Tuberculosis in South Africa and the Problem of the Native Mine Labourer' *Proceedings of the Transvaal Mine Medical Officers' Association* 22, no. 249 (1943): 174.

[66] Cartwright, *Doctors of the Mines*, 127.

[67] A. I. Girdwood, 'Tuberculosis: Examination of East Coast Recruits' *Proceedings of the Transvaal Mine Medical Officers' Association* 1, no. 5 (1921): 4.

[68] *Proceedings of the Transvaal Mine Medical Officers' Association* 1, no.5 (1921): 6.

has a much better chance in his kraal for convalescence. I consider the biggest test of Tuberculosis is the test of work. I do not think that any tuberculous native can stand six months underground.'[69]

There was, however, some dissent. Dr Frew commented that from 1918 to 1921 there had been a steady increase in both the number of cases and the number of repatriations. 'We have to consider this factor that we are sending back a great many boys who used to die on the mines, and now these boys are going back to their kraals in increasing numbers and there will be through them a rapid increase in the amount of Tuberculosis at their homes.'[70] Orenstein agreed: 'It may be said that we may have been repatriating more cases in the incipient stage, and that this is masking the death rate.'[71] The chairman, H. T. H. Butt, who had recently travelled with an experienced recruiter in the Eastern Cape, commented: 'The tuberculosis in those kraals was astonishing, distressing, not only in adults, but in children.'[72] Dr Brooke observed: 'In 1912 I was at Butterworth and during that time I examined the natives who were recruited there. The prevalence of Tuberculosis in the native territories is very great.' He found the reason in the emphasis that local communities placed on education. The schools were crowded, and the children were being infected by tubercular teachers. With less education, he said, there would be less disease.[73]

The spread of tuberculosis threatened the labour supply. As Orenstein reflected on that problem in 1921: 'Whether these boys died here and spoiled the recruiting field by upsetting their relatives at home, or whether they died at home and created a bad impression in another way and infected other people; they [the mines] were up against a serious problem in view of the fact that their reservoir of labour was drying up.'[74] The industry was also concerned about the cost of medical care and compensation. In October 1922 senior officials from the WNLA, the Departments of Native Affairs and Health, the Chamber of Mines and the SAIMR met to discuss repatriation. The meeting exposed some serious tensions between the mines and government. Speaking for the industry, Orenstein accused the government of doing nothing to halt the spread of disease. Dr Mitchell, the secretary for public health, who chaired the meeting, expressed surprise at Orenstein's charge; the government was, he said, well aware of the problem, and the issue had been raised in the House of Assembly by the minister for mines, Mr Malan.[75]

The significance of Orenstein's charge is found in the minutes of a Mine Medical Officers' Association meeting held a month later. At that meeting

---

[69] *Proceedings of the Transvaal Mine Medical Officers' Association* 1, no. 5 (1921): 5.

[70] *Proceedings of the Transvaal Mine Medical Officers' Association* 1, no. 5 (1921):10.

[71] *Proceedings of the Transvaal Mine Medical Officers' Association* 1, no. 5 (1921): 9.

[72] *Proceedings of the Transvaal Mine Medical Officers' Association* 1, no. 5 (1921): 10.

[73] *Proceedings of the Transvaal Mine Medical Officers' Association* 1, no. 5 (1921): 5, 6.

[74] *Proceedings of the Transvaal Mine Medical Officers' Association* 1, no.7 (1921): 5.

[75] Minutes of a Conference on Tuberculosis in Natives held at the Law Courts, Johannesburg, 15 October 1922. NRC 135, Miners Phthisis and Tuberculosis among Natives, Teba Archives, University of Johannesburg.

a specialist subcommittee consisting of Orenstein, Girdwood, S. Donaldson, L. G. Irvine and A. Dodds presented a report on the care of infected miners. It rejected the suggestion that men who could be cured should be treated until recovered and that those likely to die within a month or so should be detained at mine hospitals. It was adamant that the mines not provide sanatoria for blacks. Treatment on the mines was not feasible, nor was it feasible for the industry to establish clinics in rural areas.[76] The care of infected miners was the responsibility of government, not industry. Acute cases would soon have filled the hospital wards, and if men died at the mines compensation would have to be paid. There would probably be criticism from the Department of Native Affairs, and the white Mine Workers' Union (MWU) would doubtless protest about the risk to its members.

The death of miners while en route to Ressano Garcia was a constant source of bad publicity. In February 1925 Girdwood wrote a report on the repatriations. In addition to injured men, around twenty miners suffering from tuberculosis were repatriated each week. 'It is this class of case,' he wrote, 'that is so liable to die on the train. They differ from the ordinary cases inasmuch as they are suffering from an incurable disease.' It was common for miners to wait for weeks at the WNLA hospital for the Bureau to process their compensation claims. Consequently many were seriously ill by the time they were shipped out. 'If one considers the pathology of the lungs in these cases, large, ragged, breaking-down cavities full of pus, which might at any moment ulcerate through the blood vessel and cause a fatal haemorrhage, it should not be a matter of surprise that cases do die in the train, but that so few do.' Sick miners wanted to see their families before they died, and some believed a traditional healer would be able to cure them. 'If we attempted to detain all natives of this class who were not absolutely fit to travel, apart from the fact that we would very soon be overcrowded, there would be continual dissatisfaction and complaints, not only up here but in the Territories.'[77] Although the Johannesburg press carried occasional exposés of the so-called death trains during the 1920s, the system of repatriation changed little over time.

The spread of disease from the mines depended upon the policies governing 'the disposal' of migrant labour. If men were placed in care until they ceased being infectious or died, their home communities would be protected. If they were repatriated while still infectious, that was another matter. The degree of threat also depended upon the origin of the disease. If men brought tuberculosis with them to Johannesburg they would simply be returning to communities where tuberculosis was present. If recruits

---

[76] A. J. Orenstein, S. Donaldson, L. G. Irvine, A. Dodds and A. Girdwood, 'Report of the Sub-Committee of the Transvaal Mine Medical Officers' Association on Tuberculosis among Mine Natives' *Proceedings of the Mine Medical Officers' Association* 11, no. 7 (1922): 3.

[77] Letter from A. I. Girdwood, Chief Medical Officer, to General Manager WNLA, Johannesburg, 17 February 1925, WNLA 20L Diseases and Epidemics, Tuberculosis February 1923 to December 1930,Teba Archives, University of Johannesburg.

contracted tuberculosis on the mines during the 1920s their repatriation might spread infection to populations with little immunity. Those possibilities were well understood by the mine medical officers. At a Mine Medical Officers' meeting in January 1926 Donaldson asked if the WNLA was repatriating miners with infective tuberculosis. Girdwood, who was chairing the meeting, replied: 'They are repatriated from the mines; the WNLA is merely acting as agent for them.'[78] In other words, the answer was yes, but the WNLA took no responsibility.

Conclusion

There were major differences between white and black miners in terms of their health status and the work they performed. White miners enjoyed far higher rates of pay, better nutrition, better general health, better medical care and better living conditions. Because of job reservation they had lower dust exposures. Finally, if ill, they were placed in sanatorium which protected their families from infection. Black miners lived in crowded and unhygienic compounds, and once they left the mines their access to biomedical care was minimal.

The medical examinations of white and black miners were also distinct. The Miners' Phthisis Medical Bureau was responsible for the examinations of whites and the tiny number of blacks who were referred for compensation. As a result, even senior members of the Bureau knew little about black miners. Their lack of knowledge was well illustrated at a conference on tuberculosis held in Johannesburg in October 1922 and attended by E. Whitehead from the Department of Native Affairs, Watkins-Pitchford, Girdwood and Mitchell, the secretary for public health. During a heated discussion Watkins-Pitchford disagreed with his colleagues that tuberculosis on the mines was a scourge. He noted that the official incidence of 4 or 5 per 1,000 compared favourably with the rate among adult males in the industrial centres of Great Britain. Other participants in the conference pointed out that Watkins-Pitchford's figures dealt only with white miners and that many black miners who had not completed enough shifts to qualify for compensation were ill.[79]

In addition to the medical examinations, which were supposed to prevent the spread of disease and ensure the fitness of recruits, there should have been two further arms to mine medicine: improving hygiene in the compounds and the workplace and providing information to miners about the risks of dust. Kotze of the Miners' Phthisis Prevention Committee warned the Chamber as early as 1914 on the need to improve hygiene and recommended that every effort be made to identify early-stage

---

[78] *Proceedings of the Transvaal Mine Medical Officers' Association* 5, no. 7 (1926): 4.
[79] Minutes of a Conference on Tuberculosis in Natives held at the Law Courts, Johannesburg, 15 October 1922, NRC 135, Miners' Phthisis and Tuberculosis among Natives, Teba Archives, University of Johannesburg: 6.

tuberculosis. He also wanted the compounds systematically disinfected, clothing regularly washed, the walls and ceilings of compounds lime-washed, bedding boiled, floors hosed down, kitchens kept clean and underground ladder-ways systematically disinfected.[80] None of those recommendations was implemented. Two years later Watkins-Pitchford told a parliamentary select committee that tuberculosis posed a more serious danger than did silicosis because miners expectorated on their hands, and the numerous infective surfaces underground, including tools, hammers, jumpers and the sides of skips, put all classes of labour at risk. In its report for 1916 the Miners' Phthisis Prevention Committee made a number of recommendations to prevent the spread of tuberculosis, among them the disinfection of underground workings and compounds, the provision of changing-houses, the inspection of food supplies and open-air treatment for those infected.[81] Little was done.

The industry was neglectful in providing miners with information about the risks of infection. In November 1915 the director of native labour, H. S. Cooke, wrote to the Chamber asking it to issue warning notices in native languages to repatriated workers about the risks of spreading disease and said that his department was willing to provide the translations.[82] His request was ignored. Medical examinations took precedence over improvements to sanitation or information to labour about the hazards of dust and infection. The reasons were cost and control. The major barrier to prevention was the compensation system. Anna Trapido has summarised the problem as follows: 'One of the most effective disincentives to hazardous working conditions is employer cost. Failure to compensate occupational disease means that the full costs are not borne by the employers. The economic link between hazard and disease is lost.'[83]

No senior medical officer publicly criticised the system. On the contrary, without exception the senior figures in Johannesburg colluded in the practise of what can best be described as racialised medicine. In the paper they presented to the Silicosis Conference in 1930, I. G. Irvine, Anthony Mavrogordato and Hans Pirow wrote: 'The [South African] system draws around the mine natives a serviceably close net of opportunities for detection of cases of silicosis and tuberculosis.'[84] In 1931 Orenstein told a parliamentary select committee, 'I have been connected with the mines for sixteen years and I don't know of a medical organisation in the world which is more efficient than ours. To the layman it doesn't seem

[80] Letter from R. N. Kotze, Chairman Miners' Phthisis Prevention Committee, to John Munro, President of the Transvaal Chamber of Mines, Johannesburg, 26 February 1914. WNLA 173 Tuberculosis November 1913 to January 1919, Teba Archives, University of Johannesburg.

[81] *General Report of the Miners' Phthisis Prevention Committee*:17–18.

[82] Letter from H. S. Cooke, the Director of Native Labour, to the Secretary, Transvaal Chamber of Mines, 17 November 1915. WNLA 173 Tuberculosis November 1913 to January 1916. Teba Archives.

[83] Trapido, 'The Burden of Occupational Lung Disease': 73.

[84] I. G. Irvine, A. Mavrogardato and Hans Pirow, 'A Review of the History of Silicosis on the Witwatersrand Goldfields' in *Silicosis*: 203.

so if we are dealing with 1200 men a day: but we can.'[85] When Girdwood questioned the efficacy of the medical system at a MMOA meeting, Orenstein replied: 'It was very easy to say that it was possible so to arrange matters that every boy should be carefully physically examined by a medical man, but when they considered how many medical men would be necessary to do it, the proposal became ridiculous.'[86] One cost that Orenstein omitted to mention was compensation.

[85] Testimony of Dr A. J. Orenstein, Head of Sanitation Dept., Rand Mines Ltd, Tuesday 5 May 1931 before the Select Committee into the Miners' Phthisis Commission Report, AN 756-1931, SC12-31, Parliamentary Library, Cape Town: 200.
[86] *Proceedings of the Transvaal Mine Medical Officers' Association* 1, no. 7 (1921): 4.

Compensation **3**

The provision of fair benefits for men and women injured in the workplace has everywhere depended upon the vigilance of trade unions, the presence of competent medical authorities and governments that maintain some critical distance from industry.[1] On the surface South Africa, which was the first state to compensate for silicosis and tuberculosis as occupational diseases, appears to have met those criteria. Stonemasons and foundry workers in the UK facing similar risks had to wait for the Silicosis Act of 1918, which was narrower in scope and provided awards that were never as generous as those available to white miners in South Africa. In contrast to the situation in the UK, where compensation was awarded for demonstrable disability,[2] under South African law compensation was awarded for silicosis even in the absence of physical impairment. The South African legislation has often been retrospective in reach and racialised in terms of medical review and the levels of awards. Whereas workmen's compensation legislation usually covers a number of industries that share a hazard such as exposure to silica dust, asbestos fibre or lead, until 1956 the South African laws applied only to gold mines.

The laws on miners' phthisis were woven around the binary pairing of miners (whites) and native labourers (blacks) – a linguistic sleight-of-hand that in the period of reconstruction after the South African War enabled legislators to racialise the labour laws without mentioning race.[3] The technologies used in medical examinations and the methods of paying

[1] Rosner and Markowitz, *Deadly Dust*; Bradley Bowden and Beris Penrose, 'Dust, Contractors, Politics, and Silicosis: Conflicting Narratives and the Queensland Royal Commission into Miners' Phthisis, 1911' *Australian Historical Studies* 37, no.128 (2006): 89–107; Sue Morrison, *The Silicosis Experience in Scotland: Causality, Recognition, and the Impact of Legislation during the Twentieth Century* (Berlin: Lambert Academic Publishing, 2010).

[2] On the UK experience see N. J. Wikeley, *Compensation for Industrial Disease* (Aldershot: Dartmouth Publishing, 1993).

[3] Martin Chanock, 'Delict and Compensation' in *The Making of the South African Legal Culture, 1902–1936: Fear, Favour, and Prejudice* (Cambridge: Cambridge University Press, 2001): 189–96.

compensation were also racialised. White miners were examined by specialists at the Miners' Phthisis Medical Bureau and had access to private physicians. Their screening involved a clinical examination, a chest X-ray and the taking of medical and work histories. Black miners were examined by mine medical officers in theory under the Board's supervision. They were paid lump sums while whites were entitled to pensions. Access to compensation for 90 per cent of the workforce fell under the authority of mine medical officers rather than government officials.

One of the constraints facing the South African legislators was the need to placate the white trade unions. As Sir William Dalrymple told the 1930 Silicosis Conference, the Rand mines were operating under a disadvantage: 'The practical problem [of compensation] is complicated by the fact that in this country in which most of the unskilled labour is done by the native, alternative avenues of employment especially for partially disabled men are probably less easy to find than in other more fully industrialized communities with a homogenous population.'[4] White miners were organised, and legislators were forced to reach an accommodation with the powerful Mine Workers' Union (MWU).

Building the compensation system

In 1903 the Weldon Commission had identified a serious risk to all underground workers, but the minister for mines, Jan Smuts, was sympathetic to industry and played down the hazard.[5] After five years of lobbying by white miners and in the wake of a damning report by the Transvaal Mining Regulations Commission, Smuts introduced the Miners' Phthisis Allowances Act No. 34 of 1911, which provided relief for white miners and contained many of the elements that featured in subsequent law. The Act established an award process overseen by a medical board, provided care for white miners and compensation for their dependents. While its emphasis was on prevention and state regulation, it did not apply to black miners or include tuberculosis. The regulation of medical certificates, the investment of compensation funds and the examination of applicants was left to the governor general pending the introduction of comprehensive legislation.[6] That legislation took the form of the Miners' Phthisis Act No. 19 of 1912, which set out procedures for the registration of miners and a system of medical surveillance and compensation.[7]

The initial South African legislation was based on British precedent

[4] *Silicosis*: 25.
[5] Katz, *The White Death*: 159–61.
[6] Isidore Donsky, 'A History of Silicosis on the Witwatersrand Gold Mines, 1910–1946' Ph.D. diss., Rand Afrikaans University, 1993: 193–94.
[7] Matthew John Smith, 'Working in the Grave: The Development of a Health and Safety System on the Witwatersrand Gold Mines, 1900 to 1939' M.A. thesis, Rhodes University, 1993.

rather than Roman Dutch law.[8] Prior to the Workmen's Compensation Acts of 1897 and 1906, British employees were assumed to have accepted the risk of injury when undertaking work and therefore had no claim against an employer for any injury they sustained. The acts were in theory a no-fault system designed to take compensation out of the courts and improve work safety by imposing the costs on employers. Compensation was based on previous earnings, with a ceiling on lump sums of £300 for death and a weekly pension of £1 for incapacity. The legislation was confined to traumatic accidents and restricted in their application to railways, mines, factories, quarries and engineering works, thus excluding large sections of the workforce. The basis for compensation was lost wages rather than need, and there was no allowance for pain and suffering. Despite those limitations, the legislation did advance the rights of British labour.[9]

The South African Miners' Phthisis Act of 1912 provided relief for men with silicosis. Awards were made by a medical board (later a medical bureau) which had authority to make awards, impose levies on scheduled mines[10] and to examine applicants, issue medical certificates and invest funds. The act recognised two stages of silicosis, and enhanced benefits were available for those with advanced disease. Awards were based on medical reviews conducted by designated doctors. Two doctors were required to examine each white miner, and when there was disagreement a third physician, appointed by the board, adjudicated. Conflict often arose when a miner's own physician diagnosed silicosis but that opinion was rejected by the board. Black miners were examined at their place of work by mine medical officers and those diagnosed with phthisis were referred to the board for further assessment and, in a minority of cases, for compensation. Mine doctors had reason not to diagnose silicosis: it was a cost employers were anxious to avoid, and it was common for migrant workers to be repatriated without medical assessment.[11]

While the mining houses wanted government and labour to share the financial burden, the 1912 legislation placed the cost principally on employers and white miners. Because miners moved from one mine to another, the bulk of compensation was funded from a levy imposed on the industry as a whole. The employer levy was set at 5 per cent of the white miners' wages, and half of it was paid by the miners themselves. After two years, the levy was increased to 7.5 per cent, raising the employer's contribution to 5 per cent whilst the miners share remained at 2.5 per cent.[12] The impost on miners relieved employers of a portion of their liability, but it became a source of grievance between the MWU and the Chamber

---

[8] Chanock, 'Delict and Compensation': 189–96.

[9] For an overview of that history, see Wikeley, *Compensation for Industrial Disease,* and P. Weindling (ed.), *The Social History of Occupational Health* (London: Croom Helm, 1985).

[10] See Donsky, *A History of Silicosis*: 193–94.

[11] See Memo: Rand Mutual Assurance Company Insurance of Miners' Phthisis Compensation (Natives) Johannesburg 8 September 1915, WNLA 144/A Native Miners' Phthisis Insurance June 1915 to December 1915, Teba Archives, University of Johannesburg.

[12] See Donsky, *A History of Silicosis*: 195.

of Mines. After sustained lobbying the levy on miners was finally revoked by the Miners' Phthisis Act No. 40 of 1919.

The progressive nature of miners' phthisis meant that a man could leave the industry in apparent good health only later to be diagnosed with silicosis. In an interim that could stretch to several years, the mine on which he had last worked might have closed leaving behind no assets. For that reason the Compensation Fund was liable for compensating for silicosis contracted on scheduled mines not only from 1911, when the first legislation was enacted, but for some years prior to that date. That policy was a departure from the principle at the heart of the British legislation that an employer was liable for compensating a workman who contracted an occupational disease while in his service. The justification was that 'the additional burden must be looked upon in the nature of a tax, which the legislature considers should be imposed upon the industry rather than upon the community'.[13]

In addition to the board's power to levy quarterly contributions, the act provided that an actuary appointed by the minister make an annual determination of the Fund's outstanding liabilities. Those liabilities were then apportioned among the scheduled mines. The outstanding liability of each mine became payable when it closed or was removed from the schedule. If a mine closed without assets, the cost of servicing its continuing liability fell upon the remaining mines.[14]

The mines acts that ran parallel with the miners' phthisis legislation introduced a system of fines and penal provisions to enforce safety in the workplace. Under the Mines and Works Act No. 12 of 1911, any person who in contravening the act endangered the safety of a mine worker was liable to a fine of up to £250 or twelve months' hard labour. Where there was serious injury the fine rose to £500, and where death was involved the penalty was £1,000 or two years' imprisonment. The penal provisions applied to any company director or secretary as well as to the manager of a mine. Those provisions referred to traumatic injury rather than occupational disease, but the principle of a duty of care was explicit. The Miners' Phthisis Acts of 1919 and 1925 followed that lead in imposing penalties on employers who failed to prevent occupational disease.

One of the factors driving legislative change was the poverty endured by the widows of white miners. Their plight was raised in commissions of inquiry and select committee hearings and in voluminous correspondence from women appealing for assistance to the prime minister's office.[15] The major complaint was that the existing pensions were too low to sustain families already burdened with debt. A number of widows appeared before the 1916 Select Committee of Inquiry into the Working of the Miners' Phthisis Acts. Mrs Maria Albrecht, whose late husband had

---

[13] *Report of the Miners' Phthisis Commission of Enquiry 1929–30* (Pretoria: Government Printer, 1930): 6–7.

[14] Statement of the Gold Producers Committee: 33.

[15] See PM Correspondence 1/1/275 148/15/15, South African National Archives.

worked underground for twenty years, was representative of those who spoke of their hardship. For a number of years Harry Albrecht had suffered from declining health. In his final months he told his wife, 'I am feeling that I shall have to leave the mine; I am absolutely finished; I cannot climb the stopes at all.' Mrs Albrecht asked him to hold out a little longer, as they were heavily in debt. A week later he collapsed at home, and it was then that she took his papers to the Phthisis Board. She told the committee that once Harry left the mines 'he was absolutely helpless'. When asked it if would have been better if he had come out of the mine earlier, she replied, 'Yes, but I asked him to go on a little longer as we were in debt. I did not drive him down, but I wanted a little more money.'[16] She nursed her husband until his death. The pension was inadequate, and to support her two small children Mrs Albrecht did needlework. Harry Albrecht's story was one of many. At stake was the Chamber's determination to control the costs of compensation. During the period from the beginning of May 1911 to the end of July 1914, white miners and their dependents received a total of £893,922 in awards. In contrast, from 1 August 1912 to 31 July 1916 black labourers and their dependents received only £39,876. Although there were on average ten black workers for every white, the total compensation paid to whites was thirty times greater.[17]

Act No. 44 of 1916 established the Miners' Phthisis Medical Bureau and made it responsible for the clinical and radiographic examinations of white miners. The Bureau's central office was staffed by experts in lung disease who conducted medicals, including pre-employment and compensation examinations. Pre-employment, periodic and final examinations of black workers were carried out by mine medical officers. The Bureau fell under the jurisdiction of the Department of Mines, and all appointments were made by the minister.

The 1916 legislation also upgraded the compensation for silicotics and, in a major innovation, included tuberculosis as an occupational disease. Because they posed a threat to other workers, infected miners were barred from underground employment. Whites with tuberculosis received the same benefits as miners suffering from primary-stage silicosis. Black labourers also qualified for compensation payable by their employers to the Director of Native Labour. The legislation was promoted by the MWU, which, despite the lack of medical evidence, was convinced that white families were at risk of infection. Dr Andrew Watt told the parliamentary select committee of 1916 that he had examined hundreds of white miners with tuberculosis but had not seen a single case in which a man's wife or children had been infected. Dr Francis Aitken, the superintendent of the Springkell Sanatorium, agreed that it was rare for tuberculosis to spread from a white miner to his wife and children. Infection was influenced by living conditions: if conditions were good, he said, the rate of infection

---

[16] Transcripts of the Select Committee on the Working of the Miners' Phthisis Acts S.C.10-15, Session: 1915–16, AN 492, 482, 484 (Cape Town Parliamentary Library).

[17] Donsky, *A History of Silicosis*: 196.

was certain to be low.[18] For a black miner to receive an award he had first to be diagnosed by a mine medical officer and have his case forwarded to the Bureau for review. If an award was granted, payment was made by the miner's employer, rather than from an industry-wide fund, through the offices of the Department of Native Affairs. This system was augmented from 1939 onward by *ex gratia* payments – payments made as charity, not as a right – by the Chamber to destitute miners.

The vast majority of tuberculosis beneficiaries were black, and in almost every public inquiry the industry challenged the status of tuberculosis as an occupational disease. In 1921 the Chamber's George Barry, reviewing the legislation, noted: 'Contrary to preconceived ideas, employment in the gold mines at the present time does not specially increase the liability in our miners to ordinary Pulmonary Tuberculosis ... I think a strong case can be made for the exclusion of compensation for pure pulmonary tuberculosis from the Act.'[19] In their submissions to the 1931 select committee, Barry and Frans Unger, representing the Chamber, argued strongly that tuberculosis unaccompanied by silicosis should be removed from the act,[20] and the industry adopted the same position in negotiating labour contracts with colonial governments. During talks with the government of Nyasaland (now Malawi) in 1940 William Gemmill emphasised a distinction between miners' phthisis and tuberculosis. Tuberculosis, he argued, had existed in the rural areas long before blacks went into the mines. While the mines did all they could to ensure that miners' phthisis was compensated, they took a very different view towards tuberculosis, which black miners contracted while in their villages. According to Gemmill, it had always been the industry's contention that pure tuberculosis was not an occupational disease.[21]

The two-stage model of disease introduced by the Act of 1912 meant that the Miners' Phthisis Medical Bureau had to determine the point at which stage-one silicosis graded into stage-two. The Act of 1916 added some precision by offering definitions. Primary-stage silicosis required a miner to display physical signs of disease but no incapacity, while stage-two required both evidence of disease and serious incapacity. A miner could receive full compensation only when his ability to work had been impaired. While that approach was consistent with workmen's compensation law, the insidious nature of silicosis presented miners with a grim

---

[18] *Miners' Phthisis Working of Acts Parliamentary Select Committee 1916*, SC 10-15 1915 Third Report, AN 4923: 617–18, 693.

[19] Report by George Barry, Legal Advisor to the Chamber, on Miners' Phthisis Commissions (of 1919) and the Draft Act, December 1921, WNLA 144B Miners Phthisis' Act 1919, July 1919 to January 1922, Teba Archives, University of Johannesburg: 5.

[20] Mr Frans Unger and Mr George Ernest Barry representing the Transvaal Chamber of Mines, transcript of evidence before Select Committee of Inquiry into the Miners' Phthisis Commission Report, AN 756-1931 SC12-31 (Cape Town: Parliamentary Library): 85–87.

[21] William Gemmill, General Manager Tropical Areas, Note for Mr K. Lambert Hall, Secretary, Nyasaland, Northern and Southern Rhodesia Inter-territorial Conference, Salisbury, 20 December 1940, Emigrant Labour Governors Survey M2/3/19, Malawi National Archives.

choice. They could leave the mines with early-stage silicosis and receive no benefits or stay on until their health was seriously impaired in order to gain an award. To resolve that problem, the Principal Act No. 40 of 1919 introduced an ante-primary stage to the schedule. A miner could receive compensation where there was evidence of disease but no physical incapacity and therefore could seek alternative employment while still able.[22] Alongside the provision of compensation for tuberculosis that innovation distinguished the gold mines from other South African industries.

Under the Miners' Phthisis Acts Consolidation Act No. 35 of 1925, awards were based on the wages earned by a miner during the 156 shifts worked prior to his certification. Scheduled mines were required to contribute to the Compensation Fund according to a stipulated formula. Initially half of the quarterly payment was based on a silicosis rate for each mine. The remainder was divided in proportion to the aggregate miners' wages bill over the previous three months and each mine's profits.[23] Up to that time blacks had been compensated only if they were certified prior to death or discharge, providing an incentive for employers to repatriate men without certification or to detain tubercular miners in hospital until they died.[24] Under the new act blacks could be compensated if diagnosed within six months of leaving the mines. Since few miners had access to biomedical care or were aware of their rights, there was no increase in the number of awards.[25]

The 1925 act also established the Medical Board of Appeal, to which white miners or their surviving dependents could appeal the Bureau's decisions. It was a victory for the MWU, which had often accused the Bureau of unfairness in its determinations.[26] During 1915, for example, 684 white miners had presented evidence from their own doctors that they were suffering from silicosis, and the board had rejected every one.[27] The appeals system brought little change; in its first five years the Board reviewed 3,927 applications, of which only 272 were successful.[28] Black miners had no right of appeal.

Compensation for a large number of so-called prior-law beneficiaries and their dependents who had been excluded from benefits was also introduced. The law's retrospective reach was consistent with previous legislation that had caused some conflict between government and the Chamber. The Act of 1912, for example, had extended benefits to white miners who had contracted disease on a scheduled mine before the legis-

---

[22] *Report of the Commissions of Inquiry into the Working of the Miners' Phthisis Acts* (Cape Town: Government Printer, 1919): 8–9.
[23] *Report of the Miners' Phthisis Commission of Enquiry 1929–30* (Pretoria: Government Printer, 1930): 5–6.
[24] Chanock, *The Making of the South African Legal Culture*: 193.
[25] Smith 'Working in the Grave: 162.
[26] Donsky, *A History of Silicosis*: 236.
[27] See *Report of the Miners' Phthisis Board for the Year Ending 31st July, 1915* (Cape Town: Government Printer, 1916): 4.
[28] Donsky, *A History of Silicosis*: 237.

lation came into force. As benefits were increased by subsequent legislation, white prior-law beneficiaries were moved to the higher schedules.

At the Miners' Phthisis Commission of Enquiry of 1929–30 the Chamber pointed out that a number of existing beneficiaries had contracted silicosis on mines that had ceased to exist prior to the 1912 Act. Consequently, the mines responsible had never contributed to the Compensation Fund. The commission agreed that the practice was at odds with the principle that an employer was liable for a disease contracted in his service and recommended that a considerable proportion of the additional burden imposed on the Fund by the Acts of 1919 and 1925 relating to defunct mines should rightfully have fallen upon the government and not upon the mining industry.[29] Government rejected that opinion.

Barry, representing the Chamber, told the 1930 Silicosis Conference that constant changes to the legislation imposed a heavy cost on industry: 'It was impossible to alter the South African legislation without upheaval,' he noted, 'because any new benefits granted produced thousands of retrospective claimants.' One of Barry's colleagues observed that in the absence of a national insurance scheme the Miners' Phthisis Act tended to be used as a kind of poor laws for whites.[30] The Chamber raised the issue at subsequent inquiries, but it was not until 1956 that it succeeded in persuading government to charge the cost of retrospective liability to taxpayers.[31]

Despite a fall in the official disease rate with the passage of each new act, the cost of compensation continued to rise. In the period from May 1911 to 30 September 1929, £11,208,015 was paid to white miners and their dependents who, in addition, benefited from grants-in-aid, free medical care, retraining and rural resettlement schemes. From March 1911 to July 1929 black miners received £702,036.[32] That pattern persisted, and in the period to 1946 benefits to white miners and their dependents totalled £24,487,000 while black labourers and their families received just £2 million.[33]

According to Chamber estimates, by 1930 the average total payment to a married white miner was around £3,500 which after his death could rise to as much as £5,000. In 1929 there were 2,049 miners on benefits and 3,547 widow beneficiaries. In the UK compensation for occupational disease was limited to 30/- per week. After death payments for a widow and children were capped at £600. In New Zealand the pension was 35/- per week and 25/- for a single man. In New South Wales under the Workmen's Compensation Broken Hill Act, the benefits were similar, with

[29] *Report of the Miners' Phthisis Commission of Enquiry*, 6: 29.

[30] *Silicosis*: 84–85.

[31] Simons, 'Migratory Labour': 123.

[32] Smith, 'Working in the Grave': 164.

[33] *Report Miners' Phthisis Board for the Period 1st April, 1941, to 31st July, 1946* and *Report of the Silicosis Board for the Period 1st August, 1946 to 31st March 1948* (Pretoria: Government Printer, 1949).

the state contributing half of the cost.[34] While those figures suggest that the South African system was generous, the industry tended to exaggerate the financial burden. In 1929, for example, the mines received an income-tax exemption of £75,000 on their phthisis obligations. The annual levy on the scheduled mines of £800,000 was paid out of the industry's pre-tax income. In addition, government paid around £53,000 in administrative costs for the Miners' Phthisis Medical Bureau and contributed to the running of Springkell Sanatorium. During that year its contributions totalled more than £200,000.[35]

The system in operation

The administration of awards to black miners fell under the control of the Director of Native Labour rather than the medical board.[36] The Director did make some effort to protect miners from employers, and on occasion he was successful. In the months following the introduction of the Miners' Phthisis Act No. 40 of 1919, a number of claims were rejected by employers on the advice of the Rand Mutual Assurance Company. Rand Mutual argued that under the new legislation no black miner was entitled to compensation for silicosis unless he had worked underground since 31 July 1919 or for tuberculosis unless he had worked for more than a month after that date.[37] The Chamber of Mines agreed, and West Rand Consolidated Mines took a test case of a miner named Alfred Motseki to the Supreme Court.

The details of the case were as follows. Alfred Motseki had worked underground for West Rand Consolidated between 31 July 1912 and 4 July 1918. On 22 August 1919 he was diagnosed with silicosis in the ante-primary stage. Three days later the Director of Native Labour filed a compensation claim on behalf of Motseki. West Rand argued that he was not entitled to compensation because he has not been employed on a scheduled mine since the passage of the Act, but Justice Gregorowski disagreed: 'A white man has a more comprehensive benefit than the native labourer. He does not lose his right to compensation merely because the mine has fallen out of the schedule, whereas the native labourer is remediless if the mine in which he contracted silicosis no longer appears on the

---

[34] Appendix G1, Statement of the Gold Producers' Committee of the Transvaal Chamber of Mines to the Miners' Phthisis Commission 1929–1930, 10th December 1929, Johannesburg, Teba Archives: 7–9.

[35] Appendix A, Chairman of the Miners' Phthisis Board, 'Analysis and Observations on the Report of the Miners' Phthisis Commission of Enquiry, 1929–1930' presented to the Select Committee of Inquiry into the Miners' Phthisis Commission Report, AN 756-1931, SC12-31 (Cape Town: Parliamentary Library): 20.

[36] Smith, 'Working in the Grave': 183.

[37] Letter from the Legal Adviser, Rand Mutual Assurance Company, to the Secretary, WNLA 25 August 1919, WNLA 144 B July 1919 to January 1922, Teba Archives, University of Johannesburg.

list.'[38] On those grounds he upheld Motseki's right to compensation. His decision forced West Rand to make additional payments to sixty-one Mozambique miners who had been repatriated without an award.

The Department of Native Affairs was concerned that white miners were entitled to have their benefits increased as their disease worsened but black miners were not and raised the issue at a number of inquiries. By the late 1930s the Chamber's legal advisers believed that if the industry failed to address that problem government was likely to intervene. On 1 June 1939 the Gold Producers' Committee (GPC) adopted a system of *ex gratia* payments to black beneficiaries whose silicosis had progressed. On 24 November 1941 it agreed to an increase in payments and on 17 January 1944 it extended benefits to recruits from the High Commissions Territories of Basutoland, Bechuanaland and Swaziland.[39] The payments were modest, and because most migrant workers were unaware of the scheme there were few beneficiaries. Between 1939 and 1943 only 342 grants were made.[40] The correspondence shows an awareness of the hardship endured by black miners and their families. Many recruits had been in debt before they left for Johannesburg, and when they were repatriated those debts increased and they fell into poverty.[41] In March 1942 Barry, citing Girdwood, reported 'a tendency on the part of Mine Medical officers to reject Natives with considerable underground history as such natives may soon be certified to be silicotic and the responsibility for compensation rests with the employer who has last signed him for underground work.'[42] There is no evidence that he or the Chamber made any effort to remedy that injustice.

[38] Judgment by J. Gregorowski in West Rand Consolidated Mines Ltd versus The Director of Native Labour in the Supreme Court of South Africa Pretoria, November 1919, WNLA 144 B Miners Phthisis' Act 1919 July 1919 to January 1922, Teba Archives, University of Johannesburg: 5.

[39] Memo 'Ex-Gratia Grant to Natives-High Commission Territories' from John Shilling, Legal Adviser to the Chamber, to the Acting Secretary, Gold Producers' Committee, 30 October 1946, NRC 390 1&2 Miners' Phthisis Compensation for Natives 1942–1949, Teba Archives, University of Johannesburg.

[40] Letter from Dr Peter Allan, Department of Health to Secretary, Chamber of Mines, subject: Repatriation of Native Tuberculotics from the Mines to Native Territories, February 1944, Miners' Phthisis Compensation for Natives 1942–1949, NRC 390 1&2, Teba Archives, University of Johannesburg.

[41] Memo 'Payments to beneficiaries-Silicosis Act, 1946 10 December 1948,' from General Manager WNLA to the Legal Consultant, Chamber of Mines, NRC 390 1&2 Miners' Phthisis Compensation for Natives 1942–1949, Teba Archives, University of Johannesburg.

[42] Memo 'Miners' Phthisis Commission: Pooling of Native Compensation,' from G. Barry, Legal Adviser, to H. Wellbeloved, the Chamber, 26 March 1942, appended to Memo to Member of the Gold Producers' Committee 8th April 1942, NRC 390 1&2, Miners' Phthisis Compensation for Natives 1942–1949, Teba Archives, University of Johannesburg.

Monitoring by commissions of inquiry

In the period from 1903 to 1956, conflict over compensation between the MWU, the Department of Mines and the Chamber of Mines resulted in more than twenty commissions and select committees of inquiry. By contrast, although hard rock mining also dominated Australia's economy, during the same period the six Australian states appointed only seven commissions into silicosis. In the USA, where silicosis killed tens of thousands of hard rock miners and foundry workers, there was a single congressional investigation. The reason for the abundance of commissions in South Africa was the clash between MWU's demands for enhanced compensation and the industry's determination to contain costs. The commissions tended to focus upon the legislation's retrospective reach, the calculation of the levy and the status of tuberculosis as an occupational disease. The plight of black miners was a notable omission. The single page of the 1903 Weldon Commission's report devoted to black miners consists of case notes from the medical officer at the Lancaster West Gold Mine at Krugersdorp. The notes describe three fatalities, two of which were subject to post-mortem. Two other men in the set of five had spent a period either in hospital or under observation before being repatriated. Three of the men were identified as machine drillers, a task associated with high dust exposures. The brevity of the case notes and the job specialisations are significant. It is also significant that two of the five returned home, probably with infective tuberculosis.[43]

The Miners' Phthisis Commission of 1912 was chaired by Dr Van Niekerk, the mines medical inspector. His committee of seven included Irvine and Watt, who conducted a survey of white miners. It was the first South African study to use X-rays as a diagnostic tool, and not surprisingly the results were cited in subsequent inquiries. While the commission recognized the importance of miners' phthisis and tuberculosis among black labourers, it chose not to investigate their condition. In passing it did, however, make a significant admission: that white rock drillers usually had four to eight black labourers who worked two machines or more and were 'therefore not so continually exposed to the dust as the natives'.[44]

The 1919 commission of inquiry found that the medical examinations of black miners were inadequate and that black miners with tuberculosis, including men who in some cases had spent months in hospital, were being discharged without medical certificates: 'We are informed that in practice numerous cases have occurred where compensation could not be

[43] 'Medical Report on Five Cases of Silicosis among Natives,' Appendix F in *Report of the Miners' Phthisis Commission 1902–1903*.
[44] *Report of a Commission into Miners' Phthisis and Pulmonary Tuberculosis* 2: 5.

legally claimed.'[45] When a miner died in hospital it was common for the death certificate to state 'general tuberculosis' when in fact he had died from pulmonary tuberculosis.[46] Such a diagnosis prevented surviving dependents from claiming compensation. The situation with living miners was no better, and the commission found that medical examinations were poor and that there was often no exit medical as required by law.

The commission of inquiry of 1929–30, chaired by Mr James Young, was held in response to complaints by the MWU about compensation awards. The MWU claimed that the Bureau's periodical examinations were cursory and that under instructions from employers it certified a pre-arranged number of cases each year.[47] The Chamber argued that, on the contrary, the awards were so generous the industry's future was at risk. The final twenty pages of the Young report were devoted to black miners. The commission had heard testimony from the Department of Native Affairs and the Director of Native Labour for improved benefits, but in the absence of information on the circumstances of beneficiaries the commission declined to recommend any change in the legislation. Two of the commissioners dissented. W. Boshoff was particularly critical of the WNLA's medicals and questioned whether periodic examinations were in fact being carried out: 'Some of these natives died in the mines [from tuberculosis] without their having ever been examined during their working period.' He went on: 'I am convinced that the figures giving the deaths from tuberculosis among natives on the mines do not reflect the true position. Black miners were repatriated as soon as they were discovered to have tuberculosis and so the majority of deaths do not take place on the mines.' In vain he recommended that initial, periodic and final medical examinations be conducted by government medical officers.[48] His criticisms were subsequently endorsed by commissions of inquiry in 1943, 1950 and 1952.

The key witnesses who appeared before the commissions and select committees were medical specialists, and it was science that supposedly drove legislative change. For that reason perhaps the most notable feature of the 1929–30 commission was its attitude towards the science. Young dismissed the MWU's complaints about the Bureau's decisions, finding instead that applicants were always given the benefit of the doubt and that the Bureau made remarkably few mistakes. It attributed the white miners' complaints to the misleading death certificates issued by private physicians who lacked the skill to make an accurate diagnosis.[49] Such a conclusion is surprising given the ruling medical orthodoxy. The commission of

---

[45] *Report of the Commissions of Inquiry into the Working of the Miners' Phthisis Acts* (Cape Town: Government Printer, 1919): 11, 12, quote at 11.

[46] *Report of the Miners' Phthisis Commission of Enquiry*: 15.

[47] *Report of the Miners' Phthisis Commission of Enquiry, Part Two*: 83.

[48] 'Reservations by Mr W. Boshoff,' in *Report of the Miners' Phthisis Commission of Enquiry, Part Two*: 101.

[49] *Report of the Miners' Phthisis Commission of Enquiry*: 16–18.

1912 had discussed at length the difficulties in accurately diagnosing silicosis, especially in its early stages. The period between 1912 and 1929 had seen no major medical advances, and experts such as Watkins-Pitchford and Watt agreed that the use of X-rays was merely an aid and that diagnosis remained difficult. The 1929–30 commission, however, refused to acknowledge the limits to the science and deferred to the Bureau's expertise. After its report the controversy over diagnosis all but disappeared from public debate, as awards for both white and black miners came increasingly to be based solely on radiography.[50]

The Young Commission's report of 1930 was a victory for the Chamber, and in its submissions to the 1931 select committee it continued its attack on the compensation system. Unger and Barry, representing the Chamber, argued that since the gold industry was the mainstay of the national economy it was only fair that government should contribute to the cost of compensation.[51] They said that widows of white miners should receive a lump sum in place of the existing pension for life. Pensions greatly increased the mines overheads .They also, according to Barry, encouraged immorality as some widows would live with men rather than remarry in order to keep their benefits.[52] Unger and Barry told the committee that to contain the future liability it was necessary either to cut compensation payments or to greatly reduce the number of whites employed underground. Since few black miners contracted silicosis a reduction in the number of whites would reduce the number of claims.[53]

Of all the public inquiries, the commission of 1941–43, chaired by the former Chief Justice James Stratford, was the most important. Discussion of black miners occupied almost a third of its report which described a 'disquieting state of affairs in regard to compensation, medical examinations and after care'. The report was more sympathetic to the plight of migrant labour than its predecessors and supported the principle that an occupational disease was a cost of production that should be paid by industry. Stratford rejected the assumption underlying much of the legislation that because of their lower living standards blacks with silicosis did not suffer the same deprivation as whites. On the contrary, it argued that black miners so depended upon their capacity for hard labour that once disabled by lung disease they could not earn a living. Stratford noted that because of landlessness and overstocking the reserves were increasingly dependent on mine wages. While the wages of white miners had risen, those of blacks had not. If wages were below subsistence, then compen-

[50] See Jock McCulloch, 'Hiding a Pandemic: Dr G. W. H. Schepers and the Politics of Silicosis in South Africa,' *Journal of Southern African Studies* 35 (2009): 835–48.

[51] Mr Frans Unger and Mr George Ernest Barry representing the Transvaal Chamber of Mines: 85–87.

[52] Testimony of George Ernest Barry, the Transvaal Chamber of Mines, Tuesday 5th May 1931, before the Select Committee into the Miners Phthisis Commission Report AN 756-1931, SC12-31 (Cape Town: Parliamentary Library): 88.

[53] Appendix G1, Transvaal Chamber of Mines Statement on the Report of the Miners' Phthisis Commission of Enquiry, 1929–1930: 4–5.

sation awards based on wages were bound to be inadequate. The life expectancy of black miners with tuberculosis was so short that for the purposes of compensation they ought to be regarded as totally incapacitated.[54]

Among those who testified before Stratford were Dr J. Smith, chairman of the Miners' Phthisis Medical Bureau and Dr Peter Allan, the Secretary for Public Health. Dr Smith said that while the tuberculosis rate among whites had fallen there had been no parallel decline among blacks. Dr Allan told the commission that the mines produced more tuberculosis than any other industry. The Stratford commission agreed and cited the high rate of tuberculosis among repatriated workers as evidence of the inadequacy of pre-employment medical examinations. To end the injustices of the compensation system it recommended that X-rays of black miners be made compulsory and that all examinations be carried out by the Bureau rather than by mine medical officers or the WNLA.[55]

Stratford's most radical recommendations dealt with tuberculosis. If infected men returned home untreated, it said, they were a danger to their families: 'These sufferers are entitled to free medical services.' Blacks with tuberculosis should receive a pension, free board and lodging and medical care in special sanatoria supported by the Compensation Fund. Compensation should cover lost wages, wages in kind, medical care and vocational rehabilitation, and the scheduled mines should pay.[56]

When Stratford's recommendations were costed by the Government Actuary, the calculations for white miners were based on two assumptions: that less than a third would contract silicosis and that the average compensation for a miner and his dependents was £4,100. In the period from 1938 to 1941, out of 20,000 white miners there were on average 275 new cases each year. On that basis the total cost over twenty years would be £22,550,000, and the levy would rise to £2,200,000, an increase of around £800,000 per annum. In addition the mines would pay £400,000 to the Outstanding Liabilities Trust. For black miners the calculations of lost earnings were based on a monthly wage of £3 plus food and quarters set at 1s a day, coming to a monthly total of £4 10s. Blacks worked on average ten months per year, so the annual loss was £45, which over a career lasting thirteen years totalled £486. Assuming 2,000 new cases of tuberculosis and silicosis per year, the cost over twenty years would be £19,440,000 or £972,000 per annum.[57]

Those calculations were based on the official data which in the case of black miners put the silicosis rate at less than 0.2 per cent. If we apply the current rate of around 22 per cent, then the annual cost of compensation for black and white miners in 1943 would have been in excess of £300,000,000. The net profit of Anglo American and its subsidiaries in

[54] *Report of the Miners' Phthisis Acts Commission*: 3, 25–26.

[55] *Report of the Miners' Phthisis Acts Commission*: 14–17, 29.

[56] *Report of the Miners' Phthisis Acts Commission*: 25, 30.

[57] *Report of the Miners' Phthisis Acts Commission*: 30–32.

1963, the final year of Stratford's estimates, was £11,822,000.[58] If the industry had paid just a tenth of the full cost, the gold mines would have made no profit. These figures are a useful guide in estimating the extent to which the cost of lung disease was shifted onto the labour-sending communities. In rejecting the Chamber's protests that the mines could not afford to pay, Stratford concluded that if that were true the mines must be so dangerous that they should be closed.[59]

The 1943 Commission report was the high-water mark of the South African commissions. The rise to power of the National Party in 1948 saw the number of such inquiries decline. They also changed in character, and the kind of criticism made by Stratford disappeared. That did not, however, bring an end to conflict over compensation and the levy. The Pulmonary Disability Act of 1952 led to a sudden rise in the number of successful claims. In February 1954 the minister of mines, Dr A. J. R. Van Rhijn, met with the president of the Chamber, V. H. Osborn, to discuss what the latter viewed as an alarming increase in liability. Osborn found it difficult to understand how the number of certifications could have increased so dramatically, given the improvements in working conditions. He complained that many miners were being certified when there was nothing wrong with them. Fourteen mines were making a profit of less than 3/- a ton, and most of the tonnage milled by the industry as a whole came from mines earning less than 10/- a ton. Production costs had risen sharply, he said, and there was a limit to what the industry could bear. He warned that 'the effect of the pulmonary disability awards and the demands of the Mine Workers' Union might end in the mine workers having no work at all.' Van Rhijn acknowledged the sudden increase in certifications but doubted whether it was possible for him to interfere in the Bureau's decisions. The addition of the screening committee that the Chamber had demanded seemed to him likely to raise the question as to which committee was correct.[60]

In February of the following year the secretary of mines set up an inquiry chaired by W. H. Louw to review the Pulmonary Disability Act. Its report was used as the basis for negotiations between the Chamber and the Department. Louw identified an outstanding liability of £15,000,000 requiring a levy of £600,000 per year. Using the same set of data, the Chambers estimated the liability at £21,000,000 and the annual levy at £900,000.[61] Such an increase would have had a serious impact on marginal mines, and the Chamber demanded that government assume responsi-

---

[58] *Anglo American Corporation Annual Report for 1963*: 8.

[59] *Report of the Miners' Phthisis Acts Commission*: 9.

[60] Minutes of Meeting with The Honourable Minister of Mines, Dr A. J. R. Van Rhijn, 12th February, 1954: 1; WNLA 20L Diseases and Epidemics Tuberculosis March 1953–Nov. 1954, Teba Archives, University of Johannesburg.

[61] Statement by the Transvaal and Orange Free State Chamber of Mines on the Revision and Consolidation of the Silicosis Act No. 17 of 1946 12 July 1955, WNLA 20L Diseases and Epidemics Tuberculosis December 1954 to August 1955, Teba Archives, University of Johannesburg: 2.

bility for those costs. In the following year the Pulmonary Disability Act was repealed.

Compensation under high apartheid

The commissions of inquiry appointed during the 1960s and 1970s were notable for their support of the industry and its policies. The testimony of Dr Pieter Smit, Chief Group Medical officer of Goldfields Ptd Ltd, before the commission in 1964 is representative of the tone. He complained that while government was legally responsible for tuberculosis, it was the mining industry and not the government that was treating miners. The industry provided far better care and facilities than were available elsewhere in South Africa: 'In this country, Mr Chairman, I must say the liberality of diagnosis and compensation far exceed anything I have seen anywhere else on the continent and in the United Kingdom.' The commission's chairman agreed.[62]

The major legislative change during those two decades was the Occupational Diseases in Mines and Works Act (ODMWA) No. 78 of 1973. Like the previous legislation it was racially based, but it had a far wider ambit. The Act provided compensation for silicosis, asbestosis, coal workers' pneumoconiosis, obstructive airways disease and asbestos-related lung cancer. It also covered tuberculosis contracted during service or diagnosed within twelve months of a miner's last risk shift. First-degree awards covered permanent disability of more than 10 per cent but less than 40 per cent and second-degree awards for workers with an impairment of more than 40 per cent. Compared with the compensation systems in the UK and Australia, the ODMWA was unusual in offering lump-sum payments rather than pensions. Inflation and stagnant wages in the period from 1973 to 1980 saw the value of its benefits fall in real terms by as much as 60 per cent. That fall was felt by all racial groups.[63]

From 1956 a maximum dust level of 200 particles per cubic centimetre (ppcc) was introduced onto the mines. An index of average dust concentration was calculated for each mine using hundreds of individual readings. The ODMWA established a risk rating for use in calculating the levy. Under the supervision of mine management, hundreds of konimeter samples were taken at each mine and then averaged to create an air-quality index.[64] In practice that meant that 50 per cent of miners were working in dust levels considered unsafe.[65] Whatever method is used to measure dust,

---

[62] Transcript of evidence, Dr Peter Smith, Chief Group Medical Officer of Goldfields, before the Commission of Enquiry Regarding Pneumoconiosis Compensation 1964, K269 South African National Archives: 689, 694.

[63] Trapido, 'The Burden of Occupational Lung Disease': 61–62.

[64] Trapido, 'The Burden of Occupational Lung Disease': 88.

[65] See Rina King, *Silicosis in South African Gold Mines: A Study of Risk of Disease for Black Mineworkers* (TAG/WITS Sociology Research, 1985).

however, it is difficult to establish a precise relationship between dust exposure and the development of silicosis. The only adequate measure of risk is the actual disease rate.

The ODMWA received the full support of the Commission of Enquiry on Occupational Health of 1976. In its report the commission wrote: 'Although much could probably still be done about industrial health in the mining industry, there is little in the gold mining industry about which the Republic need be ashamed.' The history of the gold mines 'speaks of a sensitivity and a willingness on the part of the authorities as well as of the industry itself to accord humanitarian considerations their rightful place at all times'. The exemplary conditions on the mines were 'due to the fact that the worker's safety and health were their constant concern, and created an undeniable climate of industrial peace'.[66] The commission noted that the incidence of silicosis had become less serious and that the number of tuberculosis cases had also dropped sharply.[67] Three years later the gold mines began treating black miners for tuberculosis while they remained in risk work.

The commission of inquiry of 1981 recommended the establishment of a uniform compensation system to cover all diseases in all industries. No action was taken until the eve of majority rule, when under pressure from the powerful National Union of Mineworkers (NUM) the Occupational Diseases in Mines and Works Amendment Act (ODMWA) No. 208 of 1993 was passed. The Act was administered by the Department of Health, and compensation assessments were made by the Medical Bureau for Occupational Diseases. It transformed a race-based system into one based on wages. Because occupations, skill levels and wages in the mining industry had for decades discriminated against the majority of workers, those inequalities were incorporated into the new act.[68]

Historically, South Africa's occupational health legislation has run along two parallel paths, one covering gold mining and the other all other industries. That division is still manifest in the current law. The ODMWA No. 208 of 1993 deals solely with compensation of occupational lung diseases in miners, while lung diseases in non-miners are covered by the Compensation for Occupational Injuries and Diseases (COIDA) Act No. 130 of 1993. The two acts are equivalent in terms of their definitions of dust disease and how a diagnosis is to be made, but in all other respects the ODMWA provisions place miners and their dependents at a disadvantage.[69] The COIDA provides a pension, whereas under the ODMWA there are lump sums. If a worker dies in a factory accident his dependents

---

[66] *Report of the Commission of Enquiry on Occupational Health* (Pretoria: Government Printer, 1976): 6.

[67] *Report of the Commission of Enquiry on Occupational Health*: 7, 11.

[68] Trapido, 'The Burden of Occupational Lung Disease': 47.

[69] See Neil White, 'Is the ODMW Act Fair? A Comparison of the Occupational Diseases in Mines and Works Amendment Act, 1993 and the Compensation of Occupational Injuries and Diseases Act, 1993 with respect to compensation of Pneumoconiosis' MS, July 2004: 18.

may receive a pension under the COIDA, but if a miner dies of lung disease the ODMWA precludes such an entitlement for his widow. Lump-sum payments for physical impairment cannot be expected to support workers or their families for any length of time, and for that reason they are discouraged by the ILO. Not surprisingly, South African miners and their dependents often seek social security benefits, thereby transferring the costs of mining to taxpayers. Under the ODMWA there is no compensation for workers with a disability deemed to be less than 10 per cent and no provision for rehabilitation.

Senzeni Zokwana, who worked for many years at the President Steyn mine in Welkom, appeared as a witness before the Commission of Inquiry into Safety and Health in the Mining Industry in 1994 (The Leon Commission). He told the commission that the work underground was hard and that problems of power and control contributed to the dangerous conditions. President Steyn was a large mine that at one time had the deepest shaft on earth, and it had a massive workforce. Jan Prinsloo, who worked as a mine captain at President Steyn for more than twenty years, was responsible for five to six hundred men on each shift. If other supervisors were absent that number could rise to seventeen hundred.[70]

During Zokwana's time on President Steyn mine there was no full-time shop steward, and the only person to whom a miner could report safety concerns was the mine captain, who was already overworked. Besides, the latter's primary responsibility was output and production bonuses, which created a conflict of interest. In addition, there was little concern with health and safety. Miners were warned in advance of impending inspections by the Department of Mines. The shift boss would tell the men that there were bad areas that needed attention or that curtains had to be repaired. The culture was authoritarian. 'They [the miners] are under the constant control of management. They do not have the right to decide for themselves. They are controlled underground by the mine and again on the surface, they can even be dismissed because of breaking hostel rules.' Zokwana was confident that the right to participate in decisions about safety and the right to refuse dangerous work would reduce injuries and lung disease.[71]

Zokwana's home is in the Eastern Cape, and he told the commission of the widespread hardship among the families of miners who had been killed or disabled. Most miners, he said, suffered from tuberculosis and other lung diseases.[72] The nearest hospital with an X-ray facility was more than sixty kilometres from his village, and it did not report tuberculosis or silicosis to the authorities in Johannesburg. The roads were poor, and many men died because they could not get medical care.

[70] Interview with Jan Prinsloo, Welkom, 21 October 2005.
[71] Senzeni Zokwana, transcript of evidence given before Commission of Inquiry into Safety and Health in the Mining Industry held at Braamfontein, Johannesburg, 17th August 1994: 205, 160, 209.
[72] Senzeni Zokwana: 209.

The Mines Health and Safety Act No. 29 of 1996, passed in response to the recommendations of the Leon Commission, ushered in a new era of non-racial legislation. The new act, based on principles of co-operation and shared responsibility that were impossible under minority rule, was a major departure from the ODMWA in promoting employer, employee and government participation in the workplace.

Conclusion

The miners' phthisis and mines acts improved working conditions and created a system of medical surveillance and compensation. They also established an intimate relationship between industry and government that was to last throughout the twentieth century. In his annual report in 1975, the director of the Medical Bureau for Occupational Diseases, Dr F. J. Wiles, wrote: 'Miners are unique in that they must hold a medical certificate issued by the State. This certificate, on which a man's job depends, may at any time be withdrawn by the State.'[73] In practice, however, the state handed over control of the medical examinations of 90 per cent of the workforce to industry. There is a reference to that anomaly in a 1949 report by the Northern Rhodesian Commission of Inquiry into Silicosis. The commission recommended that the Northern Rhodesian Medical Bureau be solely responsible for certifying compensatible disease and performing all initial and other medical examinations. The alternative was for mine medical officers to be part of both the examining and the compensation system. In the commission's view that would put mine medical officers 'in the anomalous position of being, on the one hand in the service of the mine owners and, on the other, the judges between the mine owners and their employees'.[74] That ambiguity lay at the heart of the South African system. There was a second problem which had an equally catastrophic effect on work conditions. Epidemiology, as it is usually understood, played no part in producing the official data on disease. The silicosis and tuberculosis rates were the compensation rates, nothing more, nothing less. In the absence of follow-up studies the compensation rate became the most important datum used in public debate. That data measured risk.

One characteristic that South Africa shared with other national systems was the gap between legislative intent and what happened in the workplace. As Geoffrey Tweedale has shown in his history of the British asbestos industry, despite the 1931 legislation that created a state inspectorate, asbestos factories remained hazardous until they closed half a century later.[75]

---

[73] *Report of the Medical Bureau for Occupational Diseases for the Period 1st April 1974 to 31st March, 1975* (Pretoria: Government Printer, 1975): 6.

[74] *Northern Rhodesia Report of the Commission on Silicosis Legislation* (Lusaka: Government Printer, 1949): 11.

[75] Geoffrey Tweedale, *Magic Mineral to Killer Dust: Turner & Newall and the Asbestos Hazard*, 2nd edn (Oxford: Oxford University Press, 2001).

In *Miners' Lung* Arthur McIvor and Ronald Johnston tell a similar story about the British coal mines. In the period from 1850 to 1940 more than a quarter of a million miners were killed in traumatic accidents or disabled by lung disease. Progressive mines acts did reduce the number of accidents, but occupational disease continued unabated. Between 1993 and 2004, for example, British coal miners lodged more than five hundred thousand claims for bronchitis and emphysema.[76] Those injuries occurred in the context of powerful trade unions, a critical press and Labour governments. The same pattern was repeated in Scotland, where stonemasons and coal miners have suffered severely from silicosis.[77]

There are, however, important differences between South Africa and other places. It is probable that at some point in the mid-1920s the South African industry accepted that it was impossible to engineer dust out of the mines and that silicosis and tuberculosis, especially among migrant workers, was inevitable. That problem was hidden by the compensation system which produced the official data. Employers reached an accommodation with the MWU by granting relatively generous pensions and benefits to white miners. They may also have reached an agreement of sorts with the Department of Mines that the majority of the sick miners would go undiagnosed and uncompensated. While such a claim can be disputed, there are some features of the gold industry about which we can be certain. The commissions of 1930 and 1943 found that many black miners were not given exit examinations and therefore were denied compensation. Dr G. W. H. Schepers who worked as an intern at the Silicosis Medical Bureau from 1944 to 1952 protested to his superiors about the same practices. Jaine Roberts, in her 2009 study of living miners from the Eastern Cape, documents the failure to conduct exit medical examinations of 90 per cent of miners.[78] Such consistency over such a period of time is suggestive of a coherent policy.

[76] Arthur McIvor and Ronald Johnston, *Miners' Lung: A History of Dust Diseases in British Coal Mining* (London: Ashgate, 2007).

[77] See Morrison, *The Silicosis Experience in Scotland.*

[78] See Roberts, *The Hidden Epidemic.*

2. Gold mine compound, Johannesburg 1915
(*Source:* Museum Africa: Johannesburg)

4

A White Science

In Southern Africa a range of physicians saw at first hand the impact of silica dust on gold miners. They included mine medical officers who conducted pre-employment, periodic and exit medicals at the Witwatersrand Native Labour Association (WNLA) compound or at individual mines, the interns at the Miners' Phthisis Medical Bureau (MPMB) who adjudicated compensation claims, and government doctors in the labour-sending States of Basutoland, Nyasaland and Mozambique who saw the effects of mining on migrant workers. In addition there were scientists at the South African Institute of Medical Research (SAIMR) in Johannesburg who conducted research on silicosis. Finally, there were specialists in the Department of Health who treated miners at public hospitals for diseases including tuberculosis. Of all those physicians, mine medical officers were in the most ambiguous position. They were best situated to identify and therefore prevent occupational disease, but their ability to intervene was compromised by the terms of their employment.[1]

The scientific community in Johannesburg, like the various commissions of inquiry and the legislators in Pretoria, relied upon the data produced by the MPMB for understanding the risks facing gold miners. Unfortunately, the data had two serious limitations. In adjudicating compensation claims and thus producing the official disease rates, the Bureau relied upon the referrals of mine medical officers. Secondly, the system of medical surveillance generated almost no data on the post-employment health of black miners. The lack of reliable data meant that scientific and public debate was carried out in a void. That, however, did nothing to halt the creation of an orthodoxy about risk and the merits of repatriation.

In the period before 1910 the silicosis rate for all miners was probably between 23 per cent and 30 per cent.[2] After that date, according to the offi-

---

[1] For a study of this problem in the USA see Elaine Draper, *The Company Doctor: Risk, Responsibility, and Corporate Professionalism* (New York: Russell Sage, 2003).

[2] *The Prevention of Silicosis*: 235.

cial figures, the disease rates fell dramatically, to the point that by 1934–
35 there were reportedly only twenty-six cases of tuberculosis or tuberculosis with silicosis in white miners and 827 cases among blacks.[3] In its
report for 1937 the Bureau wrote:

> Judged by the standards of other communities of European stock, the
> incidence of simple tuberculosis among European miners on the Rand
> has never been excessive. The risk of contracting simple tuberculosis
> was no greater among white miners than for the general population.
> Indeed among the New Rand Miners [those who entered the mines after
> 1916] the incidence of simple tuberculosis is of the same order as in
> the Royal Air Force in Great Britain.[4]

It appeared to be a remarkable achievement.

Black recruits worked on six-to-nine-month contracts, which meant
that almost the entire workforce was replaced each year.[5] The official data
showed that the silicosis rate among whites was around fourteen times
higher than for blacks.[6] According to the Bureau's report for 1916, 'The
fact that the prevalence of simple silicosis among native employees is
comparatively slight is an argument for intermittent employment.'[7] The
decline in the tuberculosis rate between 1918 and 1935 was attributed to
the compulsory removal of infected men.[8] The Bureau acknowledged in
1937 that 'owing to the higher susceptibility of the African native, and
particularly the un-urbanized tribal native, to tuberculosis, the incidence
of simple tuberculosis and of tuberculosis associated with what is
commonly a minor degree of silicosis is decidedly higher among the native
labourers.'[9]

By 1922 the MPMB had been conducting medical examinations and
making decisions on compensation for six years. Despite its impressive
credentials, it had its critics. In October of that year senior officials from
the WNLA, the Departments of Native Affairs and Health, the Chamber of
Mines and the SAIMR met to discuss tuberculosis. Officers from Native
Affairs and Health questioned the accuracy of the official data. Many black
miners who were ill and had not completed sufficient shifts to qualify for
compensation were repatriated,[10] and once they had left the mines they
simply disappeared. Nothing came of the meeting.

[3] *The Prevention of Silicosis*: 224.
[4] *The Prevention of Silicosis*: 241.
[5] *The Prevention of Silicosis*: 242.
[6] Dr Lewis Godfrey Irvine, chairman of the Miners' Phthisis Medical Bureau, transcript of
evidence before the Select Committee into the Miners' Phthisis Commission Report, 27 April
1931 (Cape Town: Parliamentary Library): 58.
[7] *The Prevention of Silicosis*: 4.
[8] *The Prevention of Silicosis*: 25.
[9] *The Prevention of Silicosis*: 242.
[10] Minutes of a Conference on Tuberculosis in Natives held at the Law Courts, Johannesburg,
15 October 1922, Miners' Phthisis and Tuberculosis among Natives NRC 135, Teba Archives,
University of Johannesburg: 6.

The NRC's provincial superintendent at Maseru, Basutoland, wrote to the Secretary of the NRC in March 1923 that he was unable to provide as requested a report on former miners suffering from lung disease. The Corporation had no medical officer, and local government doctors had their hands full. In most cases miners lived several days' journey from the NRC camps, and even if a doctor was available it would cost £20 to conduct a single examination.[11] The district superintendent at Piet Retief, in the Eastern Cape, told much the same story. Most repatriated miners lived in remote areas. The nearest NRC medical officers were in Swaziland, and it would require a journey on horseback of two or three days to examine one man and make a report. The cost for each examination would be from £10 to £15.[12]

At a meeting of the Mine Medical Officers' Association (MMOA) in January 1926 Orenstein spoke at length about the fate of black miners. He reminded his colleagues that nobody knew how many repatriated miners died, how many recovered or whether repatriation was in their best interest. One had to assume, he said, that men sent home might infect other people. He argued that as a scientific body the Association could not address that question until it was known what happened to those men and what happened to their families and to others living in the villages.[13]

Some medical officers acknowledged the system's deficiencies. Records were not kept, and therefore when a man returned to the mines, as many did, there was no file on his service.[14] During 1926 and 1927 the Association presented a number of recommendations to the Chamber on how to overcome that problem.[15] It wanted a central bureau and the keeping of standardised records. In March 1927 Dr G. A. Riley wrote to the Chamber: 'At the present moment no accurate information is available of the term of service underground of the majority of mine natives with the exception of those of continuous service.'[16] The Association requested that fingerprinting be introduced so that accurate records could be compiled.

Girdwood, the WNLA's chief medical officer, took up the issue with the General Manager, pointing out that there was no reliable method for deter-

[11] Letter from Provincial Superintendent, NRC, Maseru, Basutoland, to the Secretary NRC, Johannesburg, 21 March 1923, Miners' Phthisis and Tuberculosis among Natives NRC 135, Teba Archives, University of Johannesburg.

[12] Letter from District Superintendent, NRC, Piet Retief, to the Secretary, NRC, Johannesburg 17 March 1923, Miners' Phthisis and Tuberculosis among Natives NRC 135, Teba Archives, University of Johannesburg.

[13] *Proceedings of the Transvaal Mine Medical Officers' Association* 5, no. 7 (1926): 4.

[14] E. H. Cluver, 'The Progress and Present Status of Industrial Hygiene in the Union of South Africa' *Journal of Industrial Hygiene* 11, no. 6 (1929): 204.

[15] See, for example, Letter from Scott Taylor, Transvaal Mine Medical Officers' Association, to General Manager, Transvaal Chamber of Mines, Johannesburg, 27 January 1926, WNLA 20L Diseases and Epidemics, Tuberculosis February 1923 to December 1930, Teba Archives, University of Johannesburg.

[16] Letter from G. A. Riley, Transvaal Mine Medical Officers' Association, to General Manager, Transvaal Chamber of Mines, Johannesburg, 6 March 1927, WNLA 20L Diseases and Epidemics, Tuberculosis February 1923 to December 1930, Teba Archives, University of Johannesburg.

mining a black miner's length of service, a factor that influenced the Bureau's decisions on compensation. He also noted that data would be useful in preventing the reengagement of men with tuberculosis who, having been rejected at one mine, would seek employment at another.[17] The General Manager disagreed; such a scheme would have required twenty to thirty European staff and ten African clerks at an annual cost of £6,000 to £7,000.[18] It was too expensive.

Some senior officials attempted to hide the problem. Dr L. G. Irvine, a powerful figure in Johannesburg, was appointed to the MPMB in 1916 and was its chairman from1926 until his retirement in 1939.[19] Addressing a parliamentary select committee in 1931, he said, 'You must not imagine that a native who leaves the mines and returns to his territories is simply ignored and lost sight of.' In rural areas a district surgeon would send a certificate if a miner had tuberculosis or silicosis. A warrant was sent, and the miner brought to Johannesburg for examination. It was possible that silicosis might develop several years after a man had left the mines. 'I imagine that natives know they can be examined,' he added.[20] Irvine knew well that medical services in rural areas were minimal or non-existent. In the Eastern Cape, for example, there were almost no X-ray facilities. Once a black miner was repatriated he vanished. There was no referral process.

The research community

South Africa's leading scientists were L. G. Irvine, Archibald Sutherland Strachan, F. W. Simson, Wilfred Watkins-Pitchford, Anthony Mavro-gordato, Spencer Lister, A. J. Orenstein and Andrew Watt of the Rand Mutual Assurance Company. All were employed either by the state or by the mining houses. With the exceptions of William Gorgas and Lyle Cummins, no one outside of that circle contributed to the South African literature. Some private practitioners did appear before select committees or commissions, but they did so on behalf of individual miners or the Mine Workers Union and their evidence had little impact. South African science was always company science, and in a racialised society it was also white science.

Both industry and the state were concerned about the high mortality rates from pneumonia and silicosis, and the brief of the SAIMR, which was established in 1912, was to investigate. Its first director was Watkins-

---

[17] Letter from A. I. Girdwood, Chief Medical Officer, to General Manager WNLA, Johannes-burg, 26 July 1927, WNLA 20L Diseases and Epidemics, Tuberculosis February 1923 to December 1930, Teba Archives, University of Johannesburg.

[18] Letter from Assistant Native Labour Adviser to General Manager WNLA, Johannesburg, 19 October 1927, WNLA 20L Diseases and Epidemics, Tuberculosis February 1923 to December 1930,Teba Archives, University of Johannesburg.

[19] 'Obituary Dr Louis Godfrey Irvine,' *South African Medical Journal* 13 April 1946: 192.

[20] Dr L. G. Irvine, transcript of evidence before Select Committee into the Miners' Phthisis Commission Report, AN 756-1931 SC12-31 (Cape Town: Parliamentary Library): 60.

Pitchford, and the WNLA provided the bulk of its funding. Most of Johannesburg's silicosis researchers were associated with the Institute's two divisions. The Research Division's primary focus was on silicosis and pneumonia, while the Routine Division carried out the diagnosis and treatment of miners and conducted medico-legal investigations. The Routine Division's funding came from services to government departments and the mines. At its foundation the SAIMR had a staff of seven: by 1926 its staff of seventy-five were conducting sixty-six thousand diagnostic investigations a year.[21]

The SAIMR epitomises the intimate relationship between the Chamber of Mines, the Department of Mines and the research community. Those three groups shared a research focus and a source of funding (mining revenue), and key researchers moved between one sector and another. The same men served on the miners' phthisis commissions and departmental committees and represented the Chamber at public inquiries. In South Africa possibly more than in Australia, Britain or the USA, it was science which defined risk and therefore the legal obligations of employers and the state.

The biomedical understanding of silicosis was fully formed before the establishment of the MPMB in 1916. The core body of knowledge was mostly inherited from Britain and Australia and embellished by the South African commissions of inquiry (in particular the Medical Commission of 1912), the SAIMR and the select parliamentary committees. The orthodoxy had a number of core elements: that silicosis is an insidious, life-threatening disease caused by exposure to silica dust; that it is difficult to diagnose and displays a lack of symmetry between the disease process and disability; that it is associated with a range of illnesses including bronchitis and emphysema and there is a particularly strong synergy between silicosis and tuberculosis; that vulnerable workers with dusted lungs will almost inevitably die from pulmonary infection; and that continued exposure to dust after diagnosis is always fatal. That body of knowledge stands up well to current biomedical understanding of the disease. Even so, almost every one of its elements eventually came under attack from the gold mining industry.

The development of a medical orthodoxy

The Weldon Commission of 1903 concluded that miners' phthisis was a life-threatening disease caused by fine silica dust, that it was almost impossible to diagnose in its early stages, and that it was common for a miner to notice symptoms only when the disease was already well advanced. Weldon found no evidence of tuberculosis among white miners, which was not surprising; the dust levels were so intense that drillers did

---

[21] Marais Malan, *In Quest of Health*: 35.

not live long enough to develop the disease.[22] Perhaps its most notable finding was that tuberculosis was common in black miners.[23] Irvine told the Commission that few black miners died in the compounds from phthisis because as soon as they became ill they returned home. Under questioning he admitted: 'There is an enormous amount of mortality [in the rural areas] which has never been investigated.' The only way to establish the mortality rate, he argued, was by conducting a follow-up study.[24]

Instead of addressing the problem of dust, the Chamber of Mines followed Irvine's suggestion and commissioned a study of tuberculosis in the labour-sending areas. In 1906 Dr G. A. Turner, the medical officer at the WNLA depot in Johannesburg, carried out a survey in southern Mozambique.[25] The first study of its kind, it was conducted at a time when conditions on the mines were at their worst. The compounds were overcrowded, and hundreds of migrant workers were dying each month from pneumonia. Turner was asked to enquire into the extent of pulmonary tuberculosis and to determine whether disease was being carried from the mines. The underlying question was where migrant workers contracted tuberculosis and how the industry should respond. Was it best to hospitalise or repatriate infected miners? Potentially the mines' continued access to labour from the tropical North hinged on the outcome.

Turner was given full co-operation by the Portuguese authorities, and he spoke at length with physicians at Lourenço Marques. They told him that tuberculosis was rife in the villages and that it was being spread by returning miners. Turner visited some of the areas from which labour was recruited and examined those who were ill, making notes on the patient's physical state, work history, symptoms and a diagnosis. He had no X-ray equipment, nor did he use sputum testing, though by his own admission a microscopic test was essential for an accurate diagnosis.[26]

Turner reported that the huts were well ventilated and nutrition was good and he found little evidence of disease: 'Tuberculosis is an exceedingly rare disease among the East Coast natives in their kraals.'[27] He supported his findings by reference to the small number of tuberculosis cases identified at the WNLA's Ressano Garcia transit station and the low incidence of infection among recruits arriving at the mines. Of the 32,682 East Coasters examined by medical officers in Johannesburg in the twelve months ending in September 1906, only thirteen were rejected because of

---

[22] *Report of the Miners' Phthisis Commission 1902–1903*: xxi–xxii, viii, ix.

[23] 'Minutes of Evidence, Dr Donald Macaulay' in *Report of the Miners' Phthisis Commission 1902–1903*: 22.

[24] 'Minutes of Evidence, Dr Louis G. Irvine,' in *Report of the Miners' Phthisis Commission 1902–1903*: 18.

[25] George A. Turner, *Report on the Prevalence of Pulmonary Tuberculosis and Allied Diseases in the Kraals of the Natives of Portugese East African Territory, South of Latitude 22°* (Johannesburg: Hayne & Gibson, 1906).

[26] Turner, *Report on the Prevalence of Tuberculosis*: 7–8.

[27] Turner, *Report on the Prevalence of Tuberculosis*: 13.

tuberculosis. He considered this to 'prove that Pulmonary Tuberculosis is comparatively rare in the kraals, and that it is a disease which is certainly not being spread throughout the country by labourers returning to the East Coast from the mines.'[28] Turner was equally sure that silicosis was rare because of the short periods men worked underground. Men returning from Johannesburg were in better physical condition than those who had never been to the mines. In the villages men lacked stamina because of their constant drinking, which in Turner's view was the reason for the high morbidity rates among recruits during the initial period of employment.[29] He concluded that the population of the East Coast was declining not because of tuberculosis but because of alcohol and venereal disease.

Turner's results were welcomed by the Chamber, and over the next thirty years they were cited routinely at commissions of inquiry, but they left a number of important questions unanswered. Tuberculosis was by Turner's admission common on the mines. Of the 187 post-mortems made of East Coasters who died at the WNLA hospital in 1905, 63 had tuberculosis. The disease was particularly acute, and Turner acknowledged that 'once infected on the Rand, [a miner] has little chance of getting back to his kraal'.[30] If there was no tuberculosis in the recruiting areas, how then could disease on the mines be explained? And why was tuberculosis not being spread to the rural areas?

Following Turner's report, conditions on the mines remained largely unchanged and the deaths continued. A sample of twenty post-mortems of recruits who arrived in Johannesburg between January and mid-April 1908 is representative of the fate of thousands of young men. All were migrant workers dispatched from Ressano Garcia, and all died at the WNLA hospital.[31] The reports are distressingly brief. In some instances, such as that of case No. 66864, the notes consist of eight or ten words. Each man has a serial number but no name. The man's age, his date of arrival, the date of hospitalisation if relevant, and the date of death are recorded. In most cases the cause of death is pneumonia, meningitis or tuberculosis. There are brief comments on external appearance (often 'poorly nourished' and in a few instances 'emaciated'). Of those who died from pneumonia case No. 71347 is typical. He arrived from Ressano Garcia on the 13 February 1908 and was admitted to hospital on 8 March, having served less than a month underground. He died on 21 March from pneumonia.

Some of the post-mortem reports are particularly disturbing. Case No. 70651 was twenty-eight when he arrived from Ressano Garcia on 14 January 1908. Admitted to the WNLA hospital on that day, he died five weeks later from tuberculosis. His general condition is described as

[28] Turner, *Report on the Prevalence of Tuberculosis*: 14.
[29] Turner, *Report on the Prevalence of Tuberculosis*: 18.
[30] Turner, *Report on the Prevalence of Tuberculosis*: 15.
[31] The original notebooks recording the post-mortem results are held at the National Institute of Occupational Health in Johannesburg by Jill Murray.

'emaciated'. The record of No. 66553, who was twenty-three, is almost identical. He arrived in Johannesburg on 13 January 1908, and having worked underground for a few weeks, he died from general tuberculosis on 28 March.

The industry's failure to act on the 1903 commission report led eventually to the Miners' Phthisis Commission of 1912 (usually known as the Medical Commission). It was chaired by Dr Van Niekerk, the mines medical inspector, and its final report contained all the elements of a medical orthodoxy. The commission agreed with Weldon that silicosis was a serious and potentially fatal disease that might progress over many years, causing frequent chest infections and attacks of pleurisy, morning coughs, shortness of breath and a susceptibility to bronchitis and heart disease. There was little correlation between the disease process and demonstrable disability. It also found that even at post-mortem it might be impossible to arrive at a correct diagnosis, and therefore the statistics on disease prevalence were unreliable. An accurate diagnosis required a clinical examination in which the physician listened carefully to the patient's chest augmented by X-rays and work and medical histories. Incapacitation often came late in the disease process, and a miner might show little loss of capacity until he ceased work. The Medical Commission noted that the association between silicosis and tuberculosis had long been recognised and that many cases of miners' phthisis ended as tuberculosis.[32] It also pointed to heart disease, emphysema and bronchitis as closely associated with exposure to silica dust.

To overcome the lack of reliable data, the Commission examined a cohort of 3,136 white miners and recorded work and medical histories. Another 326 miners were given a special examination that included an X-ray and a sputum test. Dr Andrew Watt carried out the X-rays at the Simmer and Jack Mine Hospital. The mean age of the sample was thirty-three years and almost half had had no previous occupation. Thirty-two per cent proved to have miners' phthisis, with the incidence depending upon their length of service and the nature of their work. Those at greatest risk were machine drillers, trammers and hammer men. Of those affected two-thirds had early-stage disease and an average of eight years' employment. The Commission was sure that if those men continued to work the disease would quickly progress.[33]

Although the disease was most severe among machine drillers, 48 per cent of whom were affected, the Commission cautioned that no class of underground worker was free of serious risk. The disease was a drain on skilled labour and involved great loss to families. Because silicosis was so difficult to diagnose, a board of specialists was required to make decisions on compensation. The Commission also recommended pre-employment medical examinations and early retirement so that miners in the initial stage

---

[32] *Report of a Commission into Miners' Phthisis and Pulmonary Tuberculosis*: 11, 27.
[33] *Report of a Commission into Miners' Phthisis and Pulmonary Tuberculosis*: 15, 22, 23.

of disease could take up other work. It recommended better ventilation, the provision of changing houses and improved conditions in the compounds.[34]

After 1912 medical discourse in Johannesburg became less sophisticated as X-rays replaced clinical examinations and the taking of work histories. The body of knowledge assembled by the Medical Commission did not disappear, but its reception developed some contradictory features. The most notable were repeated admissions by senior researchers in Johannesburg on the difficulty of diagnosis and the shortcomings of the mass medical reviews used at the WNLA and the industry's defence of its medical procedures. In an article published in the *Medical Journal of Australia* in 1923, Watkins-Pitchford acknowledged that, at best, medical examinations of white miners identified only half the cases of miners' phthisis in its later stages.[35] Four years later he wrote that the best diagnosis required placing a worker under observation for six months.[36] Despite those comments, at no point did he openly question the value of the WNLA's medical examinations and neither did his colleagues.

Soon after the Medical Commission tabled its final report, Turner, who had been promoted to Chief Medical Officer at the WNLA, returned once more to the question of tuberculosis. In 1913 he wrote a report on black miners in which he set out to answer three questions: what accounted for the high incidence of tuberculosis on the mines, whether the mine medical examinations were adequate and whether tuberculosis was being spread to rural areas. Relying upon records left by medical missionaries from the previous century, he suggested that tuberculosis had been common, especially in the Cape, long before the gold mines began operating and therefore the mines were not responsible for spreading disease.

Turner argued that it was poor personal hygiene and the wearing of European clothes rather than the breathing of silica dust that were the major factors.[37] When clothes became wet with rain or sweat, blacks developed chills. 'With the advance of Europeans into the country,' he wrote, 'and with the introduction of European habits and above all clothing to native populations a wave of Tuberculosis has come amongst the aboriginal inhabitants.'[38] Consequently, Christian blacks had higher tuberculosis rates than did those who adhered to a traditional way of life. Turner's 'trappings of civilisation' argument, in which colonial Africans failed to adapt to the demands of modernity, was common to imperial discourses about sexual conduct and mental illness.[39] As Randall Packard has pointed out,

[34] *Report of a Commission into Miners' Phthisis and Pulmonary Tuberculosis*: 18, 22, 23.

[35] W. Watkins-Pitchford, 'The Diagnosis of Silicosis,' *Medical Journal of Australia* 2, no. 15 (1923): 322.

[36] Watkins-Pitchford 'The Silicosis of the South African Gold Mines': 121.

[37] For a brief account of what was for a time an orthodoxy among white physicians see Packard, *White Plague, Black Labor*: 49.

[38] G. A.Turner, 'Report on Tuberculosis among Natives' undated (October 1913?), WNLA 173 Tuberculosis November 1913 to January 1916, Teba Archives, University of Johannesburg.

[39] See Jock McCulloch, *Colonial Psychiatry and the African Mind* (Cambridge: Cambridge University Press, 2005).

those who adopted European clothes were in closest contact with whites and therefore more exposed to infection.[40]

Turner's report was in part a response to the official data that showed a sharp rise in the mortality rate between 1907, when there were 501 deaths, and 1912, when there were 1,029. He acknowledged that since those figures excluded men who had been repatriated or had died at the WNLA compound, they greatly underestimated the actual number. A high proportion of deaths were among East Coasters and Tropicals (migrants from north of 22 degrees South latitude), who also accounted for 70 per cent of repatriations. East Coasters were, he recognised, among the first recruited, served longer contracts and were usually employed on machine drills. Although the recently published Medical Commission report of 1912 had commented at length on the synergy between silica exposure and tuberculosis, Turner ignored that connection. On the contrary, he claimed that silicosis was not so common as was supposed.[41]

As the death toll from pneumonia continued to mount, eventually in March 1913 the government in Pretoria, under pressure from London, barred the Chamber from further recruiting in the North. The decision was a blow for an industry with an insatiable appetite for labour. In order to regain access to Tropicals, the Chamber needed to reduce the death rate. To that end it commissioned Sir Almroth Wright and Spencer Lister in a futile attempt to find a vaccine for bacterial pneumonia.[42] It also turned to Major William C. Gorgas, who had achieved fame for his work on the Panama Canal. In December 1913 Gorgas arrived in Johannesburg to investigate the cause of so many deaths and recommend a remedy.

Gorgas was a scrupulous researcher, and his report on the human ecology of the mines was the first of its kind. In fact there was no comparable research until the end of minority rule. Gorgas found that fatalities from pneumonia varied between individual mines and according to the origin of the workers. Most cases occurred during the first months of service, and fatalities were highest among miners from the British Nyasaland Protectorate and lowest in recruits from the Cape, suggesting that immunity was a factor. Gorgas was highly critical of the mine rations, which he considered inadequate for men performing hard labour:

> I have never seen so large a proportion of the ration supplied by one article as is here supplied by mealie meal (maize). The two chief components of the daily ration are two pounds of mealie meal and 6.85 ounces of meat. This, I think a great deal too large a proportion of the carbohydrates for men doing the hard manual labour that the natives do.[43]

[40] Packard, *White Plague, Black Labor*: 49.
[41] Turner, 'Report on Tuberculosis among Natives': 8, 10.
[42] Malan, *In Quest of Health*: 95–112.
[43] *Recommendation as to Sanitation Concerning Employees of the Mines on the Rand made to the Transvaal Chamber of Mines*, W. C. Gorgas, Surgeon-General, United States Army; Chief Sanitary Officer, Isthmian Canal Commission 1914, Johannesburg: 349.

He noted that the compounds were crowded and full of litter and that miners wore wet and soiled clothes. The bucket system of waste disposal added to the unsanitary conditions.

The recommendations of Major Gorgas were modelled on his Panama experience, where the scattering of workers from crowded barracks into single huts had dramatically reduced the death rate from infectious disease: 'For the sanitation of pneumonia I would urge a similar measure on the Rand. Place your negro labourers in individual buildings, and bring in and place with them their families.' Married men living in such housing would form a permanent, skilled, and efficient workforce. Gorgas noted that while the industry was spending a million pounds a year on recruitment, the construction of family locations would save the greater part of that expense. If workers were housed in huts and provided an adequate diet, in a year or two their immunity to infection would be greatly increased. Gorgas acknowledged that because of the threat of sexual crime the white public would object to the presence of a large body of black labourers living permanently in Johannesburg.[44]

The Gorgas report was also a landmark in the knowledge of tuberculosis. It made clear that pneumonia and tuberculosis on the mines were linked by the conditions that gave rise to them – malnutrition, overcrowding and poor hygiene. For both diseases the most vulnerable populations were black migrant workers. Gorgas noted that miners' phthisis lowered a workman's resistance to tuberculosis. Transmission was usually by means of the expectorated sputa of diseased men, which, once dried and airborne, could enter the lungs of co-workers. There was a high rate of infection and during 1912 over 1,100 tuberculosis patients were repatriated by the WNLA. The mortality rate of 5.65 per 1,000, although lower than for pneumonia, was significant. Gorgas noted that in comparison the mortality rate from tuberculosis among men, women and children in London in 1911 was 1.03 per 1,000, and in New York City it was 1.67 per 1,000.[45]

Gorgas concluded that most men with miners' phthisis died from tuberculosis implanted upon a silicotic lung rather than from silicosis itself. 'Judging from findings of autopsies, I am of opinion that a considerable number of cases of deaths, reported as pneumonia, are really tuberculosis. I am inclined to think that for the future, present conditions continuing, tuberculosis will cause you more trouble among natives than does pneumonia at present.'[46] He opposed the continued use of migrant labour because of its inefficiency and human cost. As with pneumonia, he considered the most important preventive measure the replacement of the compound system. If implemented such a scheme would almost certainly have reduced the spread of tuberculosis. Gorgas also recommended regular medical examinations and the exclusion of infected men from the mines.

[44] *Recommendation as to Sanitation*: 345–47, 355, quotation at 345–46.
[45] *Recommendation as to Sanitation*: 343.
[46] *Recommendation as to Sanitation*: 343–44.

The Chamber ignored the Gorgas recommendations on labour stabilisation and compounds. In addition to the expense of providing housing for black miners and their families, stabilisation would have been unpopular with the white electorate. As a result, pneumonia remained a serious problem. According to the official data, between 1933 and 1938 it accounted for between 29.65 and 37.87 per cent of deaths from all diseases.[47]

There is no record of Turner's reaction to the Gorgas report, but it is almost certain that he was unimpressed. In January 1916 Turner gave evidence before a parliamentary select committee. By his own admission his ten years at the WNLA had given him an 'exceptional chance of forming an opinion as to the prevalence of miners' phthisis, especially as it is a practice to hold a post mortem examination on every native who dies in the compound'.[48] He told the Committee that miners' phthisis was not so prevalent as was generally supposed, suggesting that Native Affairs inspectors pushed medical officers to diagnose disease where none might exist. He also made reference to his study of rural Mozambique, which showed low rates of infection.[49] He said nothing about dust.

Three of Turner's colleagues also appeared before the Committee in 1916. Irvine reported that while white miners showed silicosis in all its stages from ante-primary to extensive fibrosis, black miners did not; he had rarely seen an advanced case of silicosis in a black miner.[50] The typical profile was of early or intermediate silicosis superimposed with a mass of tuberculosis. Black miners died quickly, he said, because they were inherently susceptible. Watkins-Pitchford explained that it was particularly difficult to diagnose phthisis in blacks: 'Very often in examining a native you find that he does not seem to be very ill and he makes no complaint. And yet you come back an hour afterwards and you find that he is dead. The native does not know when he is ill; the white man does.' He acknowledged that blacks were more likely to develop tuberculosis than whites because 'a slight degree of silicosis will render a native still more liable to develop tuberculosis'.[51] Watt told the Committee that infection among blacks on the mines was due to poor personal hygiene and a lack of immunity to infection.[52] Neither Irvine, Watkins-Pitchford nor Watt mentioned that black miners had the heaviest dust exposures, were in poor general health and lived in mine compounds.

[47] *Proceedings of the Transvaal Mine Medical Officers' Association* 19, no. 214 (1940): 229–30.
[48] Dr George Albert Turner, evidence before Miners' Phthisis Working of Acts Parliamentary Select Committee 1916, SC 10-15 1915, Third Report, AN 4923, 580. Parliamentary Library, Cape Town.
[49] Turner, evidence before Miners' Phthisis Working of Acts Parliamentary Select Committee 1916: 580–84, 589–90.
[50] Dr L. G. Irvine, evidence before Miners' Phthisis Working of Acts Parliamentary Select Committee 1916: 614.
[51] Dr Watkins-Pitchford, evidence before Miners' Phthisis Working of Acts Parliamentary Select Committee 1916: 687, 682.
[52] Dr Watt, evidence before Miners' Phthisis Working of Acts Parliamentary Select Committee 1916: 598.

Turner, who had been the first to explain miners' disease by reference to susceptibility, knew well the conditions on the mines, and yet for him tuberculosis had nothing to do with the crowded compounds, dusty stopes or malnourished recruits about which Gorgas had commented. To Turner the cause lay in the inability of black miners to cope with hard labour and the rigours of civilisation. Mine medical officers gradually adopted Turner's model with one notable change. His claim that there was no tuberculosis in rural areas was replaced with the conviction that tuberculosis was common in rural communities and migrant workers brought the disease with them to the mines.

Turner saw the villages as places of leisure where miners recuperated from brief periods of hard work. It was a powerful trope that mine doctors wedded to an image of tuberculosis spreading inexorably from contaminated rural areas to the mines. That notion, which justified the repatriation of sick miners, was more influential than the Gorgas report. The Miners' Phthisis Prevention Committee of 1916 endorsed repatriation as the best policy because it hastened the recovery of migrant workers. Besides tuberculosis was already present in districts where blacks had to some extent adopted European clothing and modes of living.[53]

Medical involution

In the decade after its foundation in 1921, the MMOA spent more time discussing tuberculosis than any other subject. In political terms, there was possibly no more important question than where black miners were contracting tuberculosis. If infection was being brought into the mines by migrant workers, the efficacy of the entry medical examinations was called into question. If tuberculosis was being contracted on the mines, then the problem was the compounds and dust, and repatriated miners were probably spreading disease to rural areas. Members of the Association soon reached a consensus that most of those with tuberculosis were diagnosed in the first two or three months underground, that the disease was usually contracted on the mines and that rates were highest among East Coasters, who were more susceptible than South African recruits. Medical officers also agreed that silicosis played little role.[54] While the obvious explanation for the low official silicosis rate lay in the cursory entry and exit examinations, the Rand doctors seized upon the short-term contracts of black miners. Supposedly, after brief periods underground, migrant workers returned home to recover, and those breaks in employment allowed their lungs to recuperate. That fiction was maintained despite the fact that silicosis progresses even after a worker has left a dusty occupation. It also

[53] *General Report of the Miners Phthisis Prevention Committee Johannesburg 15th March 1916* (Pretoria: Government Printing & Stationery Office, 1916): 17–18.

[54] See, for example, *Proceedings of the Transvaal Mine Medical Officers' Association* 5, no. 54 (1925).

went unchallenged even though labour stabilisation saw migrant workers return repeatedly to the mines in a pattern that was well established by the 1920s.[55]

The origin of infection was debated by the MMOA on a number of occasions. During a meeting in August 1921 Dr Allen commented that infection was rare among new recruits. 'I think that by far the greatest majority of the affected boys give a history of previous mine work. Therefore I assume that the infection amongst the natives with Tuberculosis – I am confining myself to East Coast natives only – is acquired on the mines.'[56] The synergy between silica exposure and tuberculosis was rarely discussed. At one of the Association's first meetings Girdwood remarked that the disease was particularly acute in recruits from Mozambique and wondered whether infection might be comparatively new to that group. He also wondered whether the length of their contracts was a factor in the severity of the disease. East Coasters worked one-year contracts while the Basuto, who were practically devoid of lung disease, usually served only four months.[57] Girdwood's question was not taken up by his colleagues.

In a paper on the detection and prevention of tuberculosis presented in September 1925, Dr W. Skaife listed the predisposing factors as tribe, physique, general condition, condition of the lungs and history of underground work.[58] During the lengthy discussion that followed there was a single reference to dust exposure: Dr Frew said that he did not take much notice of work history because he was sure that the immunity to infection built up on the mines 'more than counteracted the amount of Silicosis they [black miners] had in their chests'.[59]

Discussion of miners' phthisis and repatriation took place within strict boundaries. The members of the MMOA were industry employees, and because of the high staff turnover many of them were relatively junior. The association's founding members, Orenstein, Girdwood and Butt, were powerful figures, and in a company town like Johannesburg open dissent to their views was unusual. Following a heated discussion of repatriation at a meeting in January 1926 Dr Watkins gave notice of a motion 'that it is the considered opinion of Mine Medical Officers that a sanatorium for the treatment of tuberculous cases among the natives of the Union should be established by the Government'.[60] At the next meeting, on 18 February, he withdrew his notice.

[55] *Report of the Miners' Phthisis Medical Bureau for the Twelve Months ending July 31, 1924* (Pretoria: Government Printer, 1925): 28.

[56] *Proceedings of the Transvaal Mine Medical Officers' Association* 1, no.5 (1921): 6.

[57] Girdwood, 'Tuberculosis: Examination of East Coast Recruits': 5.

[58] W. Skaife, 'Detection and Prevention of Tuberculosis' *Proceedings of the Transvaal Mine Medical Officers' Association* 5, no. 5 (1925): 3.

[59] *Proceedings of the Transvaal Mine Medical Officers' Association* 5, no. 5 (1925): 3.

[60] *Proceedings of the Transvaal Mine Medical Officers' Association* 5, no. 7 (1926): 8.

A crisis of sorts

In February 1924 Dr Peter Allan of the Department of Health completed a report on the incidence of tuberculosis in the Transkei, the extent of resistance among the black population, the fate of repatriated miners and the effect of repatriation on the general population. At the end of the nineteenth century the Transkei, in what is now the Eastern Cape, had vast hardwood forests that local communities used for crop cultivation, grazing, hunting, the gathering of fuel and house building. Those resources had soon attracted government interest, and between 1880 and 1930 a system of state-controlled forests and areas under local authorities was established. State intervention was part of a British policy of applying scientific principles to the management of colonial landscapes.[61] By the time Allan visited the region, those policies had reshaped the local economy. The sequestering of forests helped drive men onto mines and farms, tied women to declining subsistence cultivation and contributed to the transformation of the Eastern Cape into an impoverished labour reserve.

Allan found that there was little information on the rate of infection in South Africa especially among the black majority. Lacking the resources to fill in the gaps, he relied on incomplete local hospital records and anecdotal evidence from medical practitioners. He was informed that *sifuba* (tuberculosis) was prevalent and that many children appeared to suffer from chest complaints. The data from Umtata, for example, showed that around 10 per cent of hospital admissions were tubercular. Allan traced 112 miners who had been repatriated to the Eastern Cape. Of that group 65, or 58.2 per cent, had died of tuberculosis within the first year.[62] Those results were disturbing, but Allan acknowledged that without a comprehensive survey it was impossible to gauge the extent to which the mines were spreading infection. He recommended the removal of patients to a hospital where they could be supervised.[63] He made no recommendations with regard to the Chamber of Mine's repatriation policies. His conclusion that tuberculosis was common in the Transkei and Ciskei did not absolve the mines as a source of infection.

Allan was not the first to highlight the need for further research. As director of the SAIMR, Watkins-Pitchford lobbied the Gold Producers Committee without success for funding. At a conference held at the Law Courts in Johannesburg in October 1922, he argued that there was an urgent need for a study of tuberculosis in native mine labourers.[64] When

---

[61] See Jacob A. Tropp, *Natures of Colonial Change: Environmental Relation in the Making of the Transkei* (Athens: Ohio University Press, 2006).
[62] *Report of Tuberculosis Survey of the Union of South Africa* (Cape Town: Cape Times Ltd., Government Printers, 1924): 15.
[63] *Report of Tuberculosis Survey*: 33.
[64] Minutes of a Conference on Tuberculosis in Natives held at the Law Courts, Johannesburg: 11.

he retired in 1925 there was no one at the institute to take up the issue. The new chairman, Spencer Lister, worked on pneumonia, Mavrogordato's speciality was silicosis and Pirie did research on the plague. The eleven pages that Marais Malan devotes to tuberculosis in his history of the SAIMR are indicative of the low priority given to the disease by South Africa's premier research institute.[65] It was left to a WNLA doctor stationed in Mozambique to raise the question of the threat the mines posed to rural communities.

In November 1924 Dr L. Bostock, the district manager of the WNLA at Lourenço Marques, wrote to the Chamber's general manager, William Gemmill, about the increasing number of repatriated miners arriving in an advanced state of tuberculosis: 'Apart from occasional deaths on the train, at Ressano Garcia or on the ship, a number of natives have of late reached our main camps only to die there.' He concluded that earlier diagnosis would prevent 'adverse criticism on the humanity of the present system under which Portuguese native labourers are employed on the Rand'.[66] A week later he wrote to the WNLA director in Johannesburg about the latest batch of repatriations: 'I beg to advise you that some of this week's sick rejects arrived in a very weak state, one native No. 8231 dying half an hour after arrival. Our Medical Officer there reports that some of the tubercular cases are in a very advanced stage.'[67]

The WNLA correspondence shows the same problem in Basutoland. An extract from the NRC provincial superintendent's report for March 1925 contains the following warning:

> The Government Doctors here complain of the bad state of health in which the Basuto are sent back from Johannesburg; they say in some cases they are unfit to travel, and when they come here have to go into hospital and die or have to be kept for months. They say that the mines should bear the expense and keep them there.[68]

William Gemmill referred the matter to the MMOA, and the correspondence between Bostock and the WNLA was tabled at its meeting in March 1925.

The Association's chairman Dr S. Donaldson acknowledged that the issue was important but deferred discussion until Bostock could appear in person before the MMOA. That did not happen. Instead, at the May meeting the chairman tabled a letter from Gemmill asking the Association to nominate three members to a committee that would lead the Chamber's

[65] Malan, *In Quest of Health*: 119–31.
[66] Letter from Dr, Bostock, District Manager, Lourenço Marques, to the General Manager, Transvaal Chamber of Mines, Gold Producers' Committee, 20 November 1924, cited in *Proceedings of the Transvaal Mine Medical Officers' Association* 4, no. 11 (1925): 5.
[67] Memo from L. Bostock, District Manager, to WNLA Director Johannesburg, 29 November, 1924, cited in *Proceedings of the Transvaal Mine Medical Officers' Association* 4, no. 11 (1925): 5.
[68] Extract from Monthly Report of the N.R.C. Provincial Superintendent for Basutoland cited, in *Proceedings of the Transvaal Mine Medical Officers' Association* 4, no. 11 (1925): 5.

campaign against tuberculosis.[69] In June the Chamber, with the SAIMR, formed the Tuberculosis Research Committee whose members included Watkins-Pitchford, Lister, Mavrogordato, Girdwood and Orenstein. Twelve of the seventeen members had direct links with the Chamber. The committee soon discovered there were no reliable data on prevalence or death rates. The monthly returns submitted by mine medical officers to the Department of Native Affairs, for example, could not be reconciled with hospital admissions.[70]

After four years of deliberation, in 1929 the SAIMR, in collaboration with the Chamber, commissioned a study of tuberculosis with special reference to gold miners.[71] Allan, at the time the superintendent of the Nelspruit Tuberculosis Sanatorium, was appointed principal researcher. His 1924 health survey of the Transkei and Ciskei was to form the report's centrepiece. The oversight committee was chaired by Lister and included Irvine, Watkins-Pitchford, Mavrogordato and Orenstein. It was funded by the government and the Chamber over three years with an annual budget of £6,000.[72] The Committee also recruited Lyle Cummins of the Welsh National School of Medicine as a consultant. As part of the project, in August 1929 Cummins led a group of medical researchers from the SAIMR to Mozambique to discuss the problems of miners' disease with local physicians.

Cummins reported that nearly every adult male with tuberculosis in the hospital in Lourenço Marques had been to the mines and many had been several times. According to Portuguese officials, he said, 'Tuberculosis is not common amongst the natives in their kraals. The only natives that suffer severely from Tuberculosis are those returning infected from the mines in the Transvaal. The disease is but seldom met with amongst women and children.' Cummins noted that the living conditions were much better than in the Transkei and the diet generally good. Furthermore, he reported, the Portuguese provided good medical services but he dismissed what they said about tuberculosis: 'The impression seems to prevail amongst the Portuguese Officials that they owe all the tuberculosis in their Colony to the Mines. This impression does not appear to us to be justified.'[73]

[69] *Proceedings of the Transvaal Mine Medical Officers' Association* 5, no. 1 (1925): 7.

[70] *Proceedings of the Transvaal Mine Medical Officers' Association* 6, no. 1 (1926): 3.

[71] See *Tuberculosis in South African Natives with Special Reference to the Disease amongst the Mine Labourers of the Witwatersrand* (Johannesburg: South African Institute for Medical Research, 1932).

[72] *Tuberculosis in South African Natives*: 12.

[73] Letter from S. Lyle Cummins, South African Institute for Medical Research, to Sir Spencer Lister, 24 August 1929, WNLA 20L Diseases and Epidemics Tuberculosis February 1923 to December 1930, Teba Archives, University of Johannesburg.

Conclusion

As we will see the SAIMR's report on tuberculosis which was published in 1932 helped to erase from local medical discourse the well-documented synergy between dust and tuberculosis. It was a project which industry doctors had been working on for more than twenty years. The term 'simple silicosis' has a benign origin in distinguishing between silicosis and silicosis complicated by tuberculosis. In the hands of the Rand scientists, that distinction came to have a rather different significance. In 1903 the one disease that threatened all miners was acute silicosis. After 1911 two diseases emerged: simple silicosis among whites and silico-tuberculosis among blacks. Simple silicosis in whites tended to become chronic, and many white miners survived into their fifties, although the majority in ill-health.[74] Black miners developed tuberculosis and died quickly in their villages.

At the Pan-Pacific Science Congress held in Melbourne in August 1923, Watkins- Pitchford gave a paper on miners' phthisis. His presentation to an international audience was very different to the testimony he had given to select committee hearings in Johannesburg. He told the Congress that there had been a dramatic decline in the silicosis rate on South Africa's mines and he was confident further reductions could be made. In addition to the reduced incidence there had been a change in the disease type. Silicosis was no longer life-threatening, tending slowly 'toward a natural cure', and did not increase the risk of pneumonia or bronchitis.[75] While not all of his colleagues in Johannesburg would have gone so far as to characterise silicosis as a trivial disease, like Watkins-Pitchford they did play down its synergy with tuberculosis. That shift in medical orthodoxy reframed the understanding of the risks facing black miners. It was a triumph for racial science.

Historically, silicosis and tuberculosis have occurred in two global settings, the industrial North and the colonial South. Working populations in the USA and Western Europe had recurrent exposure to the bacilli over generations and therefore some resistance to infection; those in Southern Africa did not. First generation black miners were particularly vulnerable in the unhygienic conditions on the Rand. They were exposed to silica dust while underground and to tuberculosis bacilli in crowded compounds. Once infected, they carried the disease back to rural areas where individuals developed either an active and fatal tuberculosis or a primary symptomless infection. Some of those males then became second-generation miners. When they arrived in Johannesburg, many broke down under the stress of hard labour and developed active tuberculosis, which

[74] *The Prevention of Silicosis*: 231.
[75] W.Watkins-Pitchford, 'Miners' Phthisis: Its Cause, Nature, Incidence and Prevention,' *Medical Journal of Australia* 11, no. 13 (1923): 326.

they took home.[76] The compounds were a point of transmission in a co-joined pandemic: silicosis on the mines and tuberculosis in the labour-sending areas.

[76] Malan, *In Quest of Health*: 120, 126.

5
Myth Making & the 1930 Silicosis Conference

The 1930 Silicosis Conference in Johannesburg was a pivotal moment in the global response to occupational disease. Sponsored by the ILO and the Transvaal Chamber of Mines, it brought together delegates from Canada, the UK, the USA, Australia, Italy, the Netherlands and Germany. Given the slowness of travel in 1930, it was probably the first time that many of those scientists had met. The siting of the Conference was recognition of the achievements of South Africa in terms of research, data collection and state regulation. The data included the world's largest collection of X-rays of a workforce. As the British expert Arthur J. Hall observed: 'The statistics of South Africa were so unique and of such importance too all countries which had to initiate legislation and compensation that every effort should be made to make them as clear as possible.'[1] The political context was complex. A tide of litigation was engulfing US foundries, hard rock mines and cement manufacturers, there was an asbestos crisis in the UK, a commission of inquiry into miners' phthisis had been held in South Africa the previous year and an international congress on occupational diseases was scheduled in Lyon.

The impetus for the Conference came from Dr Luigi Carozzi of the ILO, William Gemmill, the General Manager of the Transvaal Chamber of Mines, and the Chief Medical Officer for the Rand Mines Group, A. J. Orenstein. To the ILO, which had been founded in 1919, silicosis was the most important of the occupational diseases that epitomised the new hazards of industrial production. The ILO carefully monitored the research coming out of Johannesburg and took an active interest in the issues of prevention and compensation.[2] It also took a keen interest in migrant labour, especially in colonial Africa.[3] Under Carozzi's leadership it scrutinised recruitment regimes, hours of work, rates of pay and the

[1] Prof. Arthur J. Hall in discussion, in *Silicosis*: 78.
[2] *Silicosis*: 1–3.
[3] 'The ILO and Native Labour' in Antony Alcock, *History of the International Labour Organisation* (London: Macmillan, 1971): 81–98.

provision of medical care. There was a voluminous correspondence between the ILO, the Chamber, the Colonial Office and the administrations of Nyasaland, Southern Rhodesia (now Zimbabwe) and Swaziland, all of which supplied miners to the Rand.[4]

The ILO hoped that the Johannesburg Conference would foster an international network of specialists promoting workplace reform. The Chamber shared the ILO's interest in controlling silicosis and viewed the meeting as a stage on which to publicise the safety of the mines and the achievements of South African science. It also, however, had another agenda of which the ILO was unaware. The gold mines were dependent upon a flow of labour, and the 1913 ban on recruitment from north of 22 degrees South latitude had hurt the industry. The reason for the ban was pneumonia. Silicosis and tuberculosis were probably responsible for more deaths, but those deaths occurred out of sight in rural areas. The Chamber was determined to have the ban lifted, but to do so it needed to accommodate a number of its critics. The most important were the British government, which controlled access to recruits from Bechuanaland, Basutoland and Nyasaland, and the ILO. From 1928 the Chamber began to lobby for expansion into the north, pressuring the British High Commission in Pretoria and the Colonial and Dominions Offices in London for their support and presenting itself to the ILO as a champion of free labour.[5] The Johannesburg Conference was an ideal forum for promoting the Chamber's interests.

The focus of the Conference was on hard rock mining and, while the seven-hundred-page final report contains references to grinders, stonemasons and foundry workers, they are incidental. The agenda covered the medical aspects of silicosis, prevention and compensation. In his speech as chairman Sir William Dalrymple remarked that while the disease rates on the Rand had been reduced, there was no solution to miners' phthisis. He hoped that the delegates would arrive at a definition of the disease, an agreed-upon description of its stages, and consensus on its radiographic features.[6] The Conference did produce a series of recommendations that influenced the direction of research, and it also helped to establish the Rand mines as the model of workplace reform.

The agenda was divided into three sections: prevention; the clinical and pathological features of silicosis; prognosis, compensation and aftercare. Within that framework two questions were dominant: a uniform standard of diagnosis to improve medical care and compensation awards and a clarification of the relationship between tuberculosis and silicosis.[7] Although sustained discussion of compensation came only at the end of the Conference, compensation featured in much of the debate.

[4] See, for example, 'Employment of Natives on the Witwatersrand Gold Mines' CO 525/173/2 (Kew: British National Archives).
[5] Alan H. Jeeves, 'William Gemmill and South African Expansion, 1920–50' paper presented at the workshop 'The Making of Class' University of the Witwatersrand, 14 February 1987: 6.
[6] *Silicosis*: 24.
[7] Sir William Dalrymple, the Chairman's speech, in *Silicosis*: 24.

The ILO was eager to draw on the expertise of the Johannesburg research community, and to that end the Chamber provided observers for each session. It also provided a secretariat. Prior to the Conference a set of preliminary reports was circulated. The overseas reporters described their national experiences while the South African specialists reviewed particular aspects of the disease. The delegates spent two days at the Miners' Phthisis Medical Bureau, where they witnessed the assessment of compensation claims. Some delegates also visited the Witwatersrand Native Labour Association (WNLA) hospital and viewed the medical examination of black miners. Arthur Hall from the UK noted the differences between the two regimes but accepted the industry's rationale that blacks worked on short contracts whereas whites were permanent. 'The result,' he wrote, 'is that comparatively few of the natives are disabled by silicosis, and that the stringent medical machinery adopted in the case of long-term European miners is not required.'[8]

The Conference proceedings present ample evidence of uncertainty about the disease, its aetiology and the methods of diagnosis. There was vigorous debate about the clinical stage at which silicosis commences and how that moment should be identified. Orenstein remarked that the South African situation was peculiar: the price of gold was fixed, and the industry was handicapped by the attitude of legislators. 'Great weight was given in legislation to the necessity of fixing compensation for silicosis sufficiently high to stimulate prophylactic measures. The Reporters [for this conference] might consider whether silicosis could be called preventable.'[9] Mavrogordato acknowledged that power drills produced fine silica particles and that, while it was easy to remove visible dust, fine particles were more harmful and more difficult to control.[10] Dr Badham of the New South Wales Department of Public Health commented that the average exposures on the Rand of around one milligram per cubic metre suggested that the prospect of reducing the incidence of silicosis 'seemed rather hopeless'.[11] Dr Middleton, the Medical Inspector of Factories in the UK, suggested that dust counts in themselves led nowhere; variables such as the free silica content, the instrument used for measurement and the fineness of the dust made it impossible to arrive at an abstract standard that was safe.[12]

At the end of the Conference the delegates agreed on the need for research into dust and silicosis of the infective type (tuberculosis) and on the importance of initial medical examinations as a key to prevention. They recommended that competent local authorities alone should decide upon compensation and that disability claims should be adjudicated by

---

[8] Arthur J. Hall, 'Some Impressions of the International Conference on Silicosis' *The Lancet*, 20 September 1930: 656.

[9] *Silicosis*: 85.

[10] *Silicosi*: 29.

[11] *Silicosis*: 28.

[12] *Silicosis*: 30.

independent medical experts. They favoured the removal of men with infective tuberculosis from dusty workplaces and the establishment of sanatoria to treat suitable cases.[13] In addition, they agreed on the need for a universal standard of dust measurement so that the risks in different industries could be compared, and concluded that the best place to conduct such work was Johannesburg.[14]

There were no recommendations for global standards, and there was no support for external assessment of risk. The delegates offered no criticism of working conditions in South Africa or of its migrant labour system. On the contrary, they adopted South Africa's gold mines as a model of what could be achieved by an industry committed to workplace safety. In his review of the conference in *The Lancet,* Arthur Hall referred to Johannesburg as 'the mecca for silicosis researchers'. He praised South African employers and the South African state for leading the world in safety, medical care and compensation.[15]

In spite of Hall's enthusiasm, there were some dissonant voices, most notably with regard to intractable dust and the problems of diagnosis. Those tensions are well illustrated in the report by L. G. Irvine, chair of the Miners' Phthisis Medical Bureau, A. Mavrogordato, a research fellow with the South African Institute of Medical Research (SAIMR), and Hans Pirow, the Government Mining Engineer. They argued that the larger mines needed expensive mechanical ventilation but that 'if the dust content of widely scattered stopes and development ends, in the latter of which most dust is created, is to be efficiently diluted, a good deal more ventilation will be required in such places, and on some mines the cost would be prohibitive'.[16] Diagnosis was equally problematic. Dr Keith Moore from Australia commented that the dependence of South African physicians on radiography for diagnosis had led to standards that were largely arbitrary.[17] Irvine acknowledged that diagnosis required an X-ray, a thorough clinical examination and a full work and medical history,[18] omitting to mention that on the Rand none of that was done with 90 per cent of miners.

The most notable feature of the South African data presented to the delegates were the variations in disease rates between whites and blacks. According to the official returns, the silicosis rate among whites was around fourteen times higher even though, because of their job specialisations, blacks had greater dust exposures. The obvious explanation lay in the cursory examinations conducted by mine medical officers. Instead

[13] *Silicosis*: 101.

[14] Hall, 'Some Impressions': 657.

[15] Hall, 'Some Impressions': 658.

[16] I. G. Irvine, A. Mavrogordato and Hans Pirow, 'A Review of the History of Silicosis on the Witwatersrand Goldfields,' in *Silicosis,* 200.

[17] Dr Moore, Corrected Minutes of the Fifth Sitting, Monday 18th August, 1930; Minutes of the 1930 International Conference on Silicosis, South African National Archives, Pretoria, GES 1998 62/33: 5–6.

[18] *Silicosis*: 59.

Mavrogordato rehearsed the ruling orthodoxy when he told his audience that 'observation had shown that the smaller incidence of silicosis among natives, as compared with Europeans, was due to their intermittent employment.'[19] Conversely, the tuberculosis rate among black miners was far higher than for whites, and here the official explanation focused on poor hygiene in rural areas and the susceptibility of blacks to infection.[20]

The South African physician Sutherland Strachan emphasised the distinction between simple silicosis and infective silicosis in which tuberculosis was present.[21] That distinction had been used by Watkins-Pitchford at the Melbourne Conference in 1923 to mark the transformation of silicosis from an acute to a chronic disease. Both Watkins-Pitchford and Strachan believed that if tuberculosis could be avoided then silicosis became an illness of no great consequence. To a limited degree that was true for white miners, who had lower dust exposures and enjoyed enhanced nutrition and living conditions, but it was not the case for the majority of miners.

While the official data showed that the mines were safe, one delegate, at least in private, had a different view. Andrew Watt, who by 1930 had been working on miners' phthisis for almost thirty years, presented a paper to the Conference describing how dust suppression and medical surveillance had seen the acute fibrosis on the Rand mines give way to a chronic form of disease and tuberculosis, which had replaced silicosis as a major cause of death among miners. Rather than criticising the mining companies, however, Watt was conciliatory:

> The payment of compensation is a very great burden on the industry and may make the difference as to whether a low-grade mine may run at a profit or not. Therefore apart from the humanitarian aspect, the prevention of miners' phthisis is a problem of great economic importance ... The Witwatersrand has led the way in the study of this very ancient disease, because unfortunately the disease has been more prevalent here than in any part of the world, and the opportunity for research and study has been made possible by the generosity and business acumen of the leaders of the industry.[22]

Watt's paper contained nothing of significance and, not surprisingly, it was tucked away at the end of the *Proceedings*. In 1925, however, he had completed a never-to-be-published manuscript of over one hundred pages on the history of miners' phthisis, the distillation of a life-time's research, which contained path-breaking data and observations on silicosis and the

[19] *Silicosis*: 45.

[20] Irvine, Mavrogordato and Pirow, 'A Review of the History of Silicosis': 203.

[21] Dr Strachan, Corrected Minutes of the Fourth Sitting, Monday, 18th August 1930, Minutes of the 1930 International Conference on Silicosis, South African National Archives, Pretoria, GES 1998 62/33: 6.

[22] A. Watt, 'Personal Experiences of Miners' Phthisis on the Rand 1903 to 1916' in *Silicosis*: 596.

synergy between silicosis and tuberculosis. Watt argued that because free silica was hazardous even in the smallest quantities, he doubted that water alone could make hard rock mines safe. He distinguished between two forms of dust: 'income dust', the dust generated by each day's work and 'capital dust' – the dust always in circulation underground. 'Income dust' had been greatly reduced by the introduction of water sprays but sprays had no effect on 'capital dust', which Watt believed was in itself sufficient to produce silicosis. He commented that if the discovery of 'capital gas' in a colliery meant that the colliery could not be worked, the existence of 'capital dust' in a mine made it unsafe to use mine air for ventilation. Therefore improving air quality would greatly increase production costs.[23] His paper suggested that it might be impossible to reduce dust to a level at which silicosis would not occur.

Like many of his colleagues, Watt was well aware of the synergy between silicosis and tuberculosis, and he considered the threat even greater on the Rand than in hard rock mines in the USA or Australia, where labour had some immunity to infection.[24] Over a number of years Watt conducted a series of animal experiments. He found that rats inoculated with tubercle bacilli and dusted with free silica were more susceptible to tuberculosis even in the absence of demonstrable lung damage. The results were particularly significant in animals with a high natural resistance to tuberculosis.[24] Watt was sure that the small isolated silicotic areas in miners' lungs might be infective from the outset, making the synergy between silicosis and tuberculosis profound.

Reflecting on research from other parts of the continent, Watt concluded: 'Experience with British and French native troops drawn from the northern tropical districts of Africa suggests that it may be suicide for a man with a negative tuberculin reaction (that is without prior exposure to tuberculosis) to enter a phthisis-producing industry.'[25] He believed that it was impossible to reduce dust to a level at which fibrosis and hence tuberculosis would not occur. The mines were dangerous for white miners and lethal for migrant workers drawn from rural communities, especially those in the tropical North.

It is difficult to explain Watt's silence about his research. He may have been intimidated by the Chamber or by his employer, Rand Mutual. He may have been bought off though, given what is known of his character, that seems unlikely.[26] Watt was not the only delegate who remained silent about the hazards of gold mining. His animal experiments were carried out with Hans Pirow, who gave a paper at the Conference but who also remained silent. It also appears that, at least in private, Mavrogordato, who had worked with J. S. Haldane in the UK before being appointed fellow in

---

[23] Andrew Watt, 'History of Miners' Phthisis on the Rand from 1903 to 1916' MS, May 1925, Adler Medical Museum, University of the Witwatersrand: 31, 115–221.

[24] Watt, 'History of Miners' Phthisis': 17, 53.

[25] Watt, 'History of Miners' Phthisis': 130.

[26] See 'Obituary Dr Andrew Hutton Watt' *The Lancet,* 18 September 1937: 715–16.

industrial hygiene at the SAIMR in 1919, shared Watt's views.[27]

Almost certainly inspired by Watt, in 1926 Mavrogordato wrote his own 120-page review of silicosis and risk, in which he identified three key problems on the Rand: the difficulties of diagnosis, the synergy between silicosis and tuberculosis, and the intractability of the dust burden. He noted that more cases of silicosis were picked up at autopsy than during routine X-rays. There were cases in which the lungs of a miner killed in an accident would at autopsy show definite signs of fibrosis even though he was at the time of death in apparent good health.[28] Mavrogordato was an authority on that subject; in the period from 1929 to 1932 at the SAIMR he conducted hundreds of post-mortem examinations of black miners who had died suddenly.[29] In the vast majority of subjects there was evidence of silicosis and or tuberculosis. Mavrogordato also agreed with Watt that subclinical changes to lung tissue found at autopsy but invisible to radiography greatly increased the risk of infection.[30] That in turn suggested that the risk of infection for black miners was very high. He did not, however, raise any of those issues at the conference.

Politics

The Conference was an uneven political contest between the ILO and the Chamber of Mines, about which the delegates appeared unaware. They most certainly had no idea of the extent to which the Conference was about the recruitment of tropical labour. William Gemmill was a dominant figure in Johannesburg and the conference was to a large extent his creation. He regularly toured the recruiting stations and handled negotiations with Union officials and colonial administrations.[31] Under Gemmill's leadership the Chamber commissioned films and newspaper articles to counter criticism of its labour practices. Gemmill also played a key role in the creation of South Africa's industrial relations system.[32] In 1919 he served on the Low Grade Mines Commission, which was chaired by the Government Mining Engineer, Sir Robert Kotze. In the same year South Africa sent three delegates to the founding meeting of the ILO in Washington DC, and it was Gemmill who represented South African employers. In 1939 he was exiled to the WNLA's office in Southern Rhodesia as punishment for having run off with his son James's

---

[27] A. Mavrogordato, 'Contributions to the Study of Miners' Phthisis,' MS, Adler Medical Library, University of the Witwatersrand: 56. An abridged version of the paper was published under the same title as a monograph by the South African Institute of Medical Research (3, no. 19 [1926]).

[28] Mavrogordato, 'Contributions to the Study of Miners' Phthisis': 50–51.

[29] *Tuberculosis in South African Natives*: 110.

[30] Mavrogordato, 'Contributions to the Study of Miners' Phthisis': 52.

[31] See Jeeves, 'William Gemmill and South African Expansion' and Yudelman, *The Emergence of Modern South Africa*: esp. 153–55, 204–7.

[32] Yudelman, *The Emergence of Modern South Africa*: 156, 206.

fiancée. Despite that indiscretion he remained the Chamber's preferred representative in negotiating labour agreements with colonial governments and represented the industry at ILO meetings, where he defended the migrant labour system.[33]

The Chamber's most powerful body was the Gold Producers Committee (GPC). Its creation in 1922 centralised decision making in the hands of six mining houses and a small group of executives headed by Gemmill and later by his son James.[34] The GPC was supported by numerous subcommittees, including the Group Medical Officers' Committee, and it influenced every aspect of mine legislation from recruitment to safety and compensation.[35] Successive drafts of each new piece of legislation were submitted by government to the GPC for review, passed on to the Chamber's legal adviser and usually a subcommittee for comment, and then returned to the GPC for final evaluation.[36] The same process was followed with the Chamber's submissions to parliamentary select committees and commissions of inquiry. There was also voluminous correspondence between the committee and the Departments of Mines and Native Affairs.

The delegates to the Johannesburg Conference were presented with a mass of data about risk and disease rates which they had no means of evaluating. They also had no means of making an independent judgment about the migrant labour system. The Chamber was adept at presenting the mines in the best possible light and it may well have been the first employers' association to use the cinema as a propaganda tool. In 1920 African Film Productions Ltd produced a film for the Chamber titled *With the WNLA in Portuguese East Africa*. The feature, which ran for ten minutes and cost the Chamber £250, was shown in cinemas throughout South Africa.[37] The WNLA board was enthusiastic about the project: 'From a propaganda point of view such a film might be useful in removing any prevailing misconceptions as to the Association's actual methods of recruiting natives. The African Films undertook to produce the film say within four months, and to exhibit same at all the principal South African

[33] See, for example, Statement by Mr W. Gemmill (South Africa), Member of the Governing Body of the International Labour Office, made to the 123rd Session of the Governing Body, November, 1953, International Labour Office, Report of the Ad Hoc Committee on Forced Labour, NRC 469 International Labour Office 1949 to 1954. Teba Archives.

[34] Yudelman, *The Emergence of Modern South Africa*: 175.

[35] The Teba Archives at the University of Johannesburg, which hold the papers from the Chamber's two recruiting arms, the WNLA and the NRC, document the committee's influence over state regulation.

[36] In 1955, for example, the secretary for mines initiated a review of the silicosis legislation with a view to increasing compensation benefits. The Gold Producers' Committee was the leading actor in that process, and it won major concessions for the scheduled mines. See Statement by the Transvaal and Orange Free State Chamber of Mines on the Revision and Consolidation of the Silicosis Act No. 17 of 1946, 12 July 1953, WNLA 20L Diseases and Epidemics, Tuberculosis December 1954 to August 1955, Teba Archives, University of Johannesburg.

[37] See *With the WNLA in Portuguese East Africa*, WNLA 211 Cinematograph Pictures: WNLA Recruiting Operations August 1920–January 1935, Teba Archives, University of Johannesburg.

towns.'[38] *With the WNLA* tells the story of the orderly flow of migrant labour to Johannesburg and of the benefits it brought to impoverished rural communities. It attacked the ban on the recruitment of labourers from the tropical North, suggesting that it was unnecessary and was damaging rural economies. One scene shows recruits arriving at a WNLA station having spent four months walking from Lake Nyasa. Because of the ban they are rejected. The men plead their case without success. A voice-over explains: 'Many employers of labour on the Transvaal would like to employ them, but the law does not permit it; they must now walk back to their homes, but are given food and clothing. In this way the Union loses many thousands of excellent Native labourers every year.'[39]

Oscillating migration was a brutal system that reconfigured the economies of Southern Africa. Poverty was an aid to recruitment, and during much of the twentieth century the aggregate earnings of migrant workers from Basutoland were substantially higher than the country's gross domestic product. The need for food, clothing and livestock drove men onto the mines. Low wages meant that few miners achieved their economic goals, and so they returned to Johannesburg again and again. The industry resisted demands for wage rise, claiming that any increase would compromise profitability and reduce the labour supply.[40] The distance between home and the mines determined whether a man saw his family every few months or once a year. The amount of land held by a family usually decided what proportion of his working life a man spent away from home.[41] As the economies of the labour-sending areas declined, that proportion increased. By the 1950s men from the Eastern Cape served on average ten contracts, of which the majority were completed before the age of forty.[42] It appears that after that age many men were incapable of sustained physical work.

Black South Africans were reluctant to work on the mines, especially where there were alternatives. Mine wages were low, and the mines had to compete for workers against farm owners and manufacturers.[43] To meet its labour needs the gold industry set up a recruiting network that eventually extended throughout Southern Africa. The WNLA and the NRC overcame the disadvantage of low wages by offering cash advances and free travel.[44] They were successful in giving the Chamber some control over the flow of labour and preventing wage competition between the mining houses.

[38] Extract from Minutes of Meeting of the Board of Management WNLA 23 August 1920, Cinematograph Pictures: WNLA Recruiting Operations August 1920–January 1935, WNLA 211, Teba Archives, University of Johannesburg.

[39] Script, 'With the WNLA in Portuguese East Africa': 3.

[40] Glass, 'The African in the Mining Industry': 10.

[41] Francis Wilson, *Migrant Labour in South Africa: Report to the South African Council of Churches* (Johannesburg: South African Council of Churches and SPRO-CAS, 1972): 81.

[42] Simons, 'Migratory Labour': 4.

[43] Yudelman, *The Emergence of Modern South Africa*.

[44] Jeeves, *Migrant Labour in South Africa's Mining Economy*.

Transport and the capitation fees paid to colonial administrations meant that workers from the tropical North cost more than South African recruits, but those costs were offset by other gains. Labour from Mozambique was available when South African recruits remained at home during spring and early summer for planting,[45] and the lack of wage employment in Mozambique and Nyasaland meant that men would accept low wages and serve longer contracts than South Africans. While the NRC contracts were usually for less than a year, many Tropicals stayed on the mines for eighteen months.[46]

The medical examinations at the point of embarkation in Mozambique or Nyasaland and at the NRC depots within South Africa were at best cursory and, as we have seen, occasionally men who were seriously ill were forwarded to Johannesburg. While those cases irritated the GPC, they had no impact on the mines' profitability. It was important that recruits be free of tuberculosis so as to minimise the spread of infection on the mines and thereby the costs of compensation. That made recruits drawn from regions where there was little endemic tuberculosis particularly attractive. Unfortunately, those recruits were susceptible to infection.

As early as 1912 the impact of the WNLA's recruiting was obvious to colonial medical officers in Nyasaland. According to the Annual Medical Report for that year, 'Recent examinations of repatriates (from the mines) have disclosed a high percentage of tubercular disease of the lungs, and legislation will be necessary in order to prevent such an infection being introduced among a population which to a marked extent must be more or less non-immune.'[47] In addition to the WNLA's recruiting there was a steady flow of voluntary labour to the South. Wages in Nyasaland were so low that men would walk more than a thousand miles to Johannesburg to find work, in a labour flow that continued even after the ban in 1913.[48] Local doctors frequently expressed concern about the spread of disease from the mines, but it is impossible to gauge the extent of the problem. In 1915 there were sixteen reported cases of tuberculosis in Nyasaland with three deaths, but that small number is more an indication of the lack of medical services than of the disease rate.[49] None of those issues were known to the Conference delegates.

The Chamber knew how difficult and expensive it would be to reduce the risk of lung disease, and that knowledge was the key to what took place at the Conference. There was no discussion of job specialisation or of relative dust exposures. The delegates were apparently unaware of the degree to which the Johannesburg statistics were racialised. There were no data

[45] Jeeves, *Migrant Labour in South Africa's Mining Economy*: 11–12.

[46] Frederick A. Johnstone, *Race and Gold: A Study of Class Relations and Racial Discrimination in South Africa* (London: Routledge & Kegan Paul, 1976): 33.

[47] Annual Medical Report for the Year Ended 31st March 1912, Nyasaland Protectorate, Malawi National Archives: 237.

[48] Jeeves, *Migrant Labour in South Africa's Mining Economy*: 237.

[49] *Annual Medical Report on the Health & Sanitary Condition for the Year Ended 31st December 1915 of the Nyasaland Protectorate* (Zomba: Government Printer, 1915): 40.

to support the claim that black miners were protected from silicosis by oscillating migration, and there was no acknowledgment that working conditions and crowded compounds encouraged tuberculosis. The delegates left Johannesburg believing that South Africa's gold industry had brought silicosis under control. It was an illusion. Within South Africa the Chamber was able to control the critics of migrant labour and, judging from the transcripts of the various miners' phthisis commissions, it was also successful in placating senior officers within the Departments of Health, Native Affairs and Native Labour. Outside of South Africa it persuaded the colonial administrations of Nyasaland, Basutoland, Swaziland and Mozambique that the migrant labour system was beneficial.

One way to judge what happened in Johannesburg in 1930 is to imagine the consequences if Andrew Watt had spoken candidly about his research. He would have demolished the myth of safe mines promoted by the Chamber and perhaps encouraged the ILO to oppose the reintroduction of Tropical labour.[50] His data could have shifted the research agenda to the subclinical effects of silica exposure and perhaps forced the SAIMR to conduct follow-up studies in rural areas. A recognition of the complexity of the disease process could have put an end to diagnosis by X-ray alone and led to a surge in compensation awards. That in turn might have encouraged a more adequate response to occupational disease in the UK, the USA, Australia and elsewhere.

After the 1930 Conference, South African science went into a period of decline. For its part, the ILO pursued a modest reform agenda. In 1933 it revised the convention on occupational diseases and in the following year it added silicosis to the schedule of diseases for which compensation should be paid. The ILO conference in 1934 published a list of the trades and industries that posed a risk of silicosis and endorsed the importance of pre-employment and periodic examinations as a means of prevention.[51] In the following decade there were no further commissions of inquiry in South Africa, the UK or Australia. The South African data continued to mislead the research community about the hazards of hard rock mining and it may well have contributed to the invisibility of silicosis as a political issue in the USA until the 1970s.[52]

---

[50] After years of lobbying the government, the Chamber recommenced recruitment in the tropical North in 1937.

[51] *Workmen's Compensation for Silicosis*: 6.

[52] See Rosner and Markowitz, *Deadly Dust*.

6
Tuberculosis & Tropical Labour

The 1930 Silicosis Conference had publicised the safety of the mines, but it was insufficient to justify the return of Tropicals. To do so the industry needed proof that the mines would not incubate tuberculosis and spread infection to rural areas. While the Conference was in session, Dr Peter Allan was conducting fieldwork in the Ciskei and the Transkei for the South African Institute of Medical Research (SAIMR) inquiry. The report, published in 1932, was the first of its kind since Turner's in 1906, and it was to be the last in the public domain until majority rule.

The project was funded by the Chamber of Mines, and the report was written by the Tuberculosis Research Committee, whose membership was dominated by men tied professionally to the industry. The report titled *Tuberculosis in South African Natives* began with the official data showing that the tuberculosis rate in black gold miners had been falling for more than a decade: by 1930 it was reportedly down to 0.7 per cent per year.[1] Many European countries had similar rates for the same age-group, suggesting that tuberculosis was no longer a problem. The Committee agreed that the clinical profiles of African and European miners differed markedly. The most notable differences were the high prevalence in first-time black recruits, accompanied by a mortality rate of more than 70 per cent within two years of diagnosis.[2] In explaining those clinical features the Committee drew upon the 'virgin soil' theory first outlined by Turner. It is indicative of his lingering influence that a paper by Turner is included as an appendix to the report.[3] Allan's intradermal tuberculin testing in the Transkei and the Ciskei showed that men over twenty years of age were heavily tuberculised and therefore had some resistance to infection.[4] Allan

---

[1] *Tuberculosis in South African Natives*: 147.
[2] *Tuberculosis in South African Natives*: 148.
[3] G. A. Turner, 'Some Anthropological Notes on South African Native Mine Labourers' Appendix No. 1 in *Tuberculosis in South African Natives*: 302–13.
[4] The results varied regionally between 66 per cent and about 88 per cent. *Tuberculosis in South African Natives*: 207.

noted in passing that malnutrition was a major risk factor in those communities.[5]

According to *Tuberculosis in South African Natives* the disease became prevalent in South Africa only after prolonged contact with tuberculised Europeans. Although infection may have been well-tolerated under natural or tribal conditions, 'this susceptibility [of black South Africans] is fraught with extreme danger when exposure to infection is accompanied by a sudden change in occupation, food, housing and mode of life'.[6] The gold mines represented such a change. The lack of biological resistance in the African was distinct from any other risk incurred in the mining industry.[7] The presence of infected men on the mines was not due to any systemic failure of entry medical examinations.[8] Over generations communities acquired resistance through intermittent contact with infected subjects. It was the absence of such contact that made members of isolated communities prone to breakdown.

While there was evidence that repatriated miners did spread tuberculosis, it was not a serious problem. The disease was so severe in miners that most died soon after returning home. In addition the migrant labour system offered some unexpected benefits: 'The repatriation of infected miners while often exaggerated as a cause of infection in the kraals made a contribution to the endemicity of tuberculosis in the Native Territories.'[9] As in Europe and North America, the disease would burn itself out as living conditions improved and communities developed resistance.

There was at that time no effective treatment. Hopeless cases detained at the mines or sent to the Witwatersrand Native Labour Association (WNLA) hospital were given a course of tonics, good feeding, fresh air and cod liver oil. The Committee found that repatriation rather than hospitalisation was the best approach. If black miners were retained, the mortality rate would rise. Besides certain cases tended to improve rapidly when they returned to a familiar diet and the 'magnificent leisure' of village life. 'Even if sanatorium treatment were made available, such an institution would probably be intensely unpopular on account of the number of deaths that would of necessity take place there.'[10] The Committee recognised that accommodation on the mines was often unsatisfactory, with some single rooms housing more than a hundred men but it rejected Gorgas's recommendation from 1914 to house recruits in townships. Far from improving miners' health, it concluded, continuous employment would only increase the incidence of silicosis and tuberculosis.[11]

Tuberculosis in South African Natives identified two types of tubercu-

---

[5] *Tuberculosis in South African Natives*: 50.
[6] *Tuberculosis in South African Natives*: 20–21.
[7] *Tuberculosis in South African Natives*: 252.
[8] *Tuberculosis in South African Natives*: 266.
[9] *Tuberculosis in South African Natives*: 248, 235, 280.
[10] *Tuberculosis in South African Natives*: 109, 252–53, 112.
[11] *Tuberculosis in South African Natives*: 126, 78.

losis in black miners, one that approximated to disease in the 'civilized' and the other, which was characteristic of Africans in general, that had a parallel in European children and adolescents. The types corresponded roughly to the categories of 'simple tuberculosis' and 'tuberculo-silicosis'. Most recruits who fell ill soon after their arrival on the Rand had definite but unrecognisable lesions that entry medicals could not pick up. On starting work they broke down.[12] The Committee's illustrious consultant, Lyle Cummins, offered an explanation as to why. He argued that just as some infected but healthy European children developed acute tuberculosis after a bout of bronchitis or on quitting the leisures of childhood, so too did a proportion of mine recruits shortly after starting work. 'It is not to be wondered that latent tuberculous infection should tend to light-up into activity when a life of monotonous leisure is suddenly exchanged for one of strenuous and unfamiliar exertion.' Most 'simple tuberculosis' was found in infected men whose lesions were 'larval'. Such recruits were 'a menace to white miners'. Here, unexpectedly, instead of referring to the racial deficits favoured by Turner, Cummins quoted at length from the US researcher Dr Leroy Gardner of Saranac Laboratories. Gardner had demonstrated that exposing guinea pigs already infected with low-grade tuberculosis to silica dust often led to a generalisation of the disease. Cummins identified the same 'acquired liability' among black miners; in a certain proportion of cases inhaled dust might light up quiescent foci of infection. In summary, the problem was exposure to silica dust. Cummins' contribution, which echoed the work of Andrew Watt, was probably the report's most important finding, but it sat uncomfortably with the body of the document.[13]

Tuberculosis in South African Natives had two narrative structures which at various points collided. It is likely that Allan, as the only full-time appointment, drafted the report which was then rewritten by Irvine, Girdwood and Orenstein. That would account for the admissions sprinkled throughout the document which run counter to its findings: references to malnutrition among recruits, the impact of exposure to dust and gruelling labour on miners' health, dramatic changes in heat and cold in moving between the surface and the stopes, and overcrowded living conditions. The problem of countervailing evidence is most obvious in the statements on dust. The Committee noted that black miners had the highest dust exposures but that whites had far higher silicosis rates. It simultaneously argued that dust levels had been reduced to such an extent that silicosis had become 'a relatively benign fibrosis'.[14] In the hands of Orenstein, Irvine and Girdwood the concept of susceptibility was racialised, thereby shifting the problem of infection away from dust in the mines and malnutrition in rural areas to immutable racial deficits.

In his survey of the Transkei from 1924 Allan had found malnutrition

---

[12] *Tuberculosis in South African Natives*: 254.
[13] *Tuberculosis in South African Natives*: 255, 263, 288, 274.
[14] *Tuberculosis in South African Natives*: 287.

a significant factor in the spread of infection. That theme was ignored in *Tuberculosis in South African Natives*.[15] The context for that decision is provided by the Eastern Cape famines of 1912 and 1927, in which few people starved to death but public health declined as growing numbers of men, women and children lived on a diet consisting mostly of carbohydrates. The historian Diana Wylie refers to such chronic malnutrition as 'starving on a full stomach' and she identifies the cause as the lack of land, and mine wages that were insufficient to sustain a family.[16]

What Wylie terms the 'malnutrition syndrome' first entered political discourse in the 1930s to explain why blacks so readily fell victim to tuberculosis. According to public health authorities, African families were poorly nourished because mothers and wives were incapable of preparing or using food properly. That problem was compounded by the careless overstocking of land, which damaged the soil and reduced incomes. The South African medical community promoted the orthodoxy, and it was used by governments to characterise black poverty as a cultural trait rather than being the result of its own policies.[17] There is an affinity between the malnutrition syndrome and the concept of susceptibility used in *Tuberculosis in South African Natives*. In combination they erased the two most important vectors of tuberculosis, namely malnutrition and silica dust.

The 1930s was a period of growth for the industry and number of black miners increased from 214,155 in 1931 to 359,710 in 1941. There was, however, little investment in the compounds. While the newer facilities housed between sixteen and twenty men in a single room, the older compounds still held as many as forty. In the majority of cases the sleeping places consisted of hard, cold concrete berths separated by concrete partitions.[18] In addition to the rise in numbers, the exclusion of experienced workers who exhibited any evidence of fibrosis helped to hide the growing pool of infection. In 1935 the WNLA revised the guidelines for the examination of recruits. Medical officers were instructed to pay particular attention to any signs of tuberculosis or silicosis. Men who had been repatriated were not to be recruited. Doctors were also to be alert to pleurisy, bronchitis and asthma. They were cautioned: 'Special care should be taken in the examination of natives with long underground mining histories.'[19] Doctors were to look for heart lesions and flabby muscles indicating wasting: 'It has been the experience that the majority of such natives, especially the older boys with a mining history, develop tuberculosis and it is useless to send such men to the mines as they will invariably be rejected by the Mine Medical officer.'[20] There was no

---

[15] *Report of Tuberculosis Survey*: 1–2.

[16] Wylie, *Starving on a Full Stomach*: 9–10, 89.

[17] Wylie, *Starving on a Full Stomach*: 198–99, 239.

[18] *Report of the Commission on the Remuneration and Conditions*: 31.

[19] Suggestions relating to the Medical Examination of Native Mine Labourers (as amended for Tropical Natives), WNLA Circular, Johannesburg, August 1935, Malawi National Archives Emigrant Labour 1935–1936 M2/3/3: 1, 2.

[20] Suggestions relating to the Medical Examination of Native Mine Labourers.

suggestion that medical officers should refer such cases to the Bureau for compensation.

The more rigorous guidelines had two important outcomes. They prevented the employment of men with obvious lung disease and they ensured that experienced miners with early stage silicosis or tuberculosis did not receive compensation. G. E. Barry, the Chamber's legal adviser, was particularly critical of the procedures: 'There is a tendency on the part of Mine Medical officers to reject Natives with considerable underground history as such natives may soon be certified to be silicotic and the responsibility for compensation rests with the employer who has last signed him for underground work.'[21] The guidelines did nothing to help mine doctors better identify early-stage disease.[22]

The invisibility of the actual disease burden was ensured by the absence of follow-up studies. During the 1940s there are a number of references in the Chamber's correspondence and in the proceedings of the Mine Medical Officers' Association (MMOA) to the lack of data on black miners once they left the industry. In March 1943 Dr A. Millar told an Association meeting that 'no statement can be made regarding the prevalence of tuberculosis' in black miners.[23] He attributed the unreliability of the existing data to the paucity of medical services in rural areas and the failure of local doctors to report cases.

Recruiting labour in Nyasaland

The lack of industry in Nyasaland (now Malawi) meant that local workers were willing to accept the wages offered in the South. According to official estimates, by the mid-1930s at least 120,000 Nyasa were working in South Africa, the Rhodesias, Tanganyika (now the mainland part of Tanzania) and Mozambique. Most of those men were employed on mines.[24] Although the number of recruits Nyasaland supplied to the Rand was far smaller than Basutoland or Mozambique, the impact on public health was significant. The experimental reintroduction of Tropicals from Nyasaland and Northern Rhodesia (now Zambia) began in 1934. Initially the numbers were modest, and by 1937 there were only 1,330 recruits. In that year Tropicals accounted for just 1 per cent of all black workers employed by the Chamber of Mines. Ten years later they accounted for 11

[21] Memo from G. Barry, Legal Adviser to H. Wellbeloved, The Chamber, 26 March 1942, appended to Memo to Member of the Gold Producers' Committee 8th April 1942 subject: Miners' Phthisis Commission, Pooling of Native Compensation, NRC 390 1 & 2 Miners' Phthisis Compensation for Natives 1942–1949, Teba Archives, University of Johannesburg.

[22] Memo from G. Barry, Legal Adviser to H. Wellbeloved.

[23] A. Millar, 'Brief Notes on the Prevalence of Tuberculosis in South Africa, with Special Reference to Native Mine Labourers' *Proceedings of the Transvaal Mine Medical Officers' Association* 22, no 249 (1943): 167.

[24] *Annual Medical & Sanitary Report for the Year Ending 31st December 1936, Nyasaland Protectorate.* (Zomba: Nyasaland, 1937): 31.

per cent, and in 1956 that had risen to 18 per cent.[25] As the number of Tropicals rose, the number of South African recruits fell. There are no records of how many Nyasa returned each year from South Africa with tuberculosis, but the data on migrant workers who had worked in Southern Rhodesia (now Zimbabwe), although incomplete, were disturbing. During the period 1932 to 1936, out of a total of 611 miners, 64 or 10.4 per cent were repatriated to Nyasaland with tuberculosis. According to the Annual Medical Report for 1936: 'The control of this disease is difficult and it is obvious that unless measures are instituted there is considerable danger of this disease proving a real menace in Nyasaland.'[26]

Soon after the ban was lifted, Gemmill wrote an assessment of the mines' labour needs. He was certain that the demand for labour would rise and that the mines' needs could not be met from within South Africa. The industry's future was at stake and it was 'absolutely essential' that the ban be permanently rescinded.[27] Gemmill acknowledged that until Tropicals became acclimatised there would be high mortality rates from pneumonia and tuberculosis.

In 1936 the Chamber commissioned a review of public health in Nyasaland, seeking insight into the health status of recruits and in particular the incidence of tuberculosis. The medical officer at Dedza identified malaria, trypanosomiasis (sleeping sickness), bilharzia and parasitic infections as major problems, but there was a very low rate of tuberculosis: 'In the Cholo District over a period of two years in which 10,000 natives would be coming for treatment, there were not more than ten natives seen with signs of active tuberculosis of any kind.' The same was true of Port Herald and the Lower Shire District: 'On village surveys performed by Medical Officers in 1934/5, in which man, woman and child were all examined from villages dotted over Nyasaland, there were only two cases recorded of tuberculosis.' In some villages tuberculosis was seen occasionally in old mine workers or their relatives. 'One can in fact say that the spread has in practically all cases been brought in from outside.'[28]

The issue of Tropicals was discussed regularly by the Gold Producers Committee (GPC) and, in a confidential memo from March 1937, Orenstein reflected on the future of recruiting. He was concerned that the mortality rates, which were higher than for Union workers, might result in the ban's being reimposed. Where large numbers were employed, the mortality rates were initially high but would, he said, gradually subside. In the previous twenty-five years the rates on the Rand had fallen from

[25] Wilson, *Migrant Labour in South Africa*: 131.

[26] *Annual Medical & Sanitary Report for the Year Ending 31st December 1936, Nyasaland Protectorate*. (Zomba: Nyasaland, 1937): 14, 15.

[27] W. Gemmill, General Manager, Transvaal Chamber of Mines, 'Native Labour,' 20 January 1936, Malawi National Archives Emigrant Labour 1935–1936 M2/3/: 1–2.

[28] 'An Examination of Natives for Work on the Rand Mines at Johannesburg and a Few Factors Bearing on the Question' the Medical Officer, Dedza, Nyasaland, 20 January 1936, Malawi National Archives Emigrant Labour 1935–1936 M2/3/3: 1–3.

around 40 to 7 per 1,000, and the same pattern was evident on the mines of the Union Miniére in the Congo. After 1930 economic conditions in the Congo saw the employment of fewer Tropicals, and those who remained became acclimatised. Whereas in 1910 the mortality rate was 100 per 1,000, in 1930 it had fallen to 16.28. While Orenstein conceded that the data were based on small numbers, he considered it 'unreasonable to assert that the mortality rate of tropical natives on the Witwatersrand gold mines is likely to be too high to justify their employment. Comparison with those presently employed natives is scarcely a fair one.'[29] It would be more useful, he argued, to compare the mine data with the rates for men of a similar age in their home communities. He expected that they might be much the same, that is, around 20 per 1,000.

The view from Nyasaland was very different. Dr C. H. Hunt, the medical officer in charge of the African Hospital at Zomba, wrote a brief report on tuberculosis in June 1937. The incidence was increasing yearly, with the number of known cases having risen from 73 in 1928 to 243 in 1935. He believed that the figures represented only a fraction of the actual numbers, and he was certain that such an increase in so short a time could not be due to improvements in diagnosis or the increasing confidence of patients in seeking medical care: 'The main source of infection lies in Rhodesia and the Union, the disease being brought back to this country by Nyasaland natives who have been employed in the South. In this way multiple foci of infection now exist in the Protectorate itself.'[30] Employers made no attempt to treat migrant workers, and they returned home to spread infection. The views of Hunt and his colleagues had no impact and the ban on recruitment was permanently lifted. Soon afterward the WNLA began negotiations with the governments of Northern Rhodesia, Nyasaland and Southern Rhodesia for an increase in recruitment. Colonial administrations were eager to reap the financial benefits of granting licences to the WNLA. While it was in their financial interest to cooperate in the flow of labour to the Rand, it was a bargain with the devil.

Money for miners

The government of Nyasaland wanted to exert some control over recruitment, but it had no capacity to monitor the health of returning miners. Its discussion with the WNLA about the health of recruits was confined to the mortality data, and there was little reference to the long-term effects of fibrosis. The negotiations were in any case markedly unequal. Gemmill

[29] Confidential Memo, A. J. Orenstein, Rand Mines Ltd., to General Manager, Transvaal Chamber of Mines, 3 March 1937, WNLA 17/1 Mortality Data Dec 1922 to March 1937, Teba Archives, University of Johannesburg: 2.

[30] Memo, Tuberculosis, Dr C. H. Hunt, the Medical Officer in Charge, African Hospital, Zomba, to the Hon. Director of Medical Services, Zomba, 2nd June 1937, Tuberculosis July 1934–Sept. 1946 Medical Reports, M2/14/3, Malawi National Archives.

often reminded colonial administrators that the gold mines could, if they chose, get their labour from Mozambique or employ Tropicals who arrived in Johannesburg of their own volition. Under those circumstances Nyasaland would receive no financial benefit and it would lose all control over the movement of migrant workers. The negotiations took place in a setting in which competition for labour was intense. White settlers in Northern Rhodesia and Nyasaland wanted access to local workers, while the gold and asbestos mines of Southern Rhodesia competed with the Rand for a limited labour supply. The negotiations also involved the Secretary of State in London and the ILO, whose conventions on the treatment of migrant workers operated in Southern Africa.

William Gemmill believed the mining industry had two alternatives:[31] either to engage Tropicals who arrived on their own account in Johannesburg or to recruit in the Territories under conditions agreed to by colonial governments. The former method was simpler and cheaper, but the Chamber of Mines took the position that regulated recruitment would be best. In reaching an agreement, a series of meetings between the Governor of Nyasaland , Sir Harold B. Kittermaster, the Chief Secretary and Gemmill were held in Zomba in February 1938. After protracted discussions Nyasaland agreed that the WNLA could recruit up to eight thousand five hundred men. It was a major increase which Kittermaster needed to justify to London. Many workers found their way to South Africa on their own initiative, and Kittermaster cautioned head office that unless an attempt was made to meet the WNLA's demands there would be an uncontrolled flow of labour to the South.[32] There were also financial considerations. The annual interest to Nyasaland on deferred pay for eight thousand five hundred recruits was around £4,000. In addition the WNLA agreed to pay a work pass or capitation fee of six pence per month per man.[33] Although that figure was well below the agreement Kittermaster had reached with Southern Rhodesian employers, the WNLA would pay the recruits' travel expenses, and the contract went ahead. In the following months the WNLA established its headquarters at Dowa, with depots at Blantyre, Bilila, Mlengeni, Lilongwe and Salima. The capital outlay was £30,000.[34]

The Chamber was pleased with the contract, and in December 1938 the WNLA requested that the quota be increased to fifteen thousand. In support of its proposal Gemmill submitted data showing a reduction in

[31] William Gemmill, Confidential Memorandum Setting Out the Association's Case for an Increase in Its Nyasaland Quota from 8,500 to 15,000 Natives, Witwatersrand Native Labour Association, Ltd., Tropical Areas Administration, Salisbury, 10th January 1942, WNLA Correspondence LB 1/5/1, Malawi National Archives: 5.

[32] Letter Confidential from Harold B. Kittermaster, Governor, Zomba, Nyasaland, to the Rt. Hon. W. Ormsby Gore, Secretary of State for the Colonies, 26th February, 1938, WNLA Correspondence LA 1/5/1, Malawi National Archives.

[33] These details are taken from a confidential dispatch from Sir Harold Kittermaster to the Secretary of State of 5 March 1938. Letter from the Labour Department, The Secretariat, Zomba, Nyasaland, 20 April 1947, M2/24/30, Malawi National Archives.

[34] Gemmill, Confidential Memorandum: 3.

deaths from disease. For the period January to October 1938 the rate was 9.78 per 1,000.[35] Although higher than the 7.10 per 1,000 for all migrant workers, Gemmill pointed out that over 80 per cent of the Nyasa were novices, which necessarily meant a higher death rate than with seasoned labour.[36] Kittermaster was satisfied and asked the Colonial Office to support a quota of ten thousand for the coming year.

The authorities in Southern Rhodesia took a very different view. They conceded that great efforts were being made to reduce occupational disease on the Rand, but they did not accept that the mortality rate was a true indication of the risks. 'For example, silicosis, even if accompanied by early tuberculosis, will not usually cause death within the contract period. It may be merely latent on discharge and later may have serious results.' The Southern Rhodesian government opposed the movement of Tropical labour to the south claiming that it would put its own population at risk. 'It is thought that sufficient time has not yet elapsed, nor is sufficient data available at present, to support an assumption that tropical Natives can so be employed without danger to their health and that of their families.'[37] No altruism was involved; the Southern Rhodesian mines and farms wanted the Nyasa who were going to South Africa.

Gemmill had little success in Northern Rhodesia. In October 1938 he met the governor, J. A. Maybin, in Lusaka to discuss increasing the quota of Barotse from the existing one thousand five hundred.[38] Maybin noted that in December 1937 recruitment had been suspended because of high mortality rates and the general shortage of labour. Gemmill conceded that while the rate for Northern Rhodesians had remained constant at around 22.75 per 1,000, he hoped to reduce it to 19. He asked permission to recruit five thousand Barotse commencing in September 1939. Maybin refused because of the number of deaths. It was a brave decision, given that capitation fees were a significant source of revenue. In 1940, for example, the government in Mozambique entered into an agreement with the WNLA to increase the number of recruits to South Africa's gold and coal mines to one hundred thousand. An annual fee of £1. 14s. 6d. per man was paid by the mines, and an annual fee of 10s. was paid by each recruit to the Mozambique administration.[39]

The outbreak of World War II saw the gold mines face an acute labour shortage. In January 1942 Gemmill, in the confidential memo quoted

[35] Letter from Harold B. Kittermaster, Governor, Zomba, Nyasaland, to the Colonial Office, 9 December, 1938, WNLA Correspondence LB 1/5/1, Malawi National Archives.
[36] Appended Nyasaland Natives Employed on the Witwatersrand Gold Mines Death Rates, Letter from Harold B. Kittermaster.
[37] Migrant Labour Agreement Memorandum, Salisbury, 6th December, 1938, WNLA Recruiting Secretariat, Chief Secretary, S1/49/38, Malawi National Archives.
[38] Note of an Interview between His Excellency the Governor (J. A. Maybin, Esq. CMG) and the General Manager, the Transvaal Chamber of Mines (W. Gemmill, Esq.) Held at Lusaka on the 27th October, 1938, WNLA Recruiting Secretariat, Chief Secretary, S1/49/38, Malawi National Archives.
[39] *Report of the Commission on the Remuneration and Conditions*: 3.

earlier, set out for the Nyasaland government the WNLA's case for an increase in its quota.[40] He argued that the gold industry was of great importance to the war effort and it could only operate with an adequate labour supply. The mines urgently needed at least forty thousand Tropicals. He pointed out that the system of recruitment conformed in every respect with the ILO convention: the gold mines offered average earnings of 2s. 2d. a day plus food, accommodation and medical care. On their return home, recruits were usually in much better physical condition than when they left. The medical services were of a high standard, and miners who showed signs of pulmonary disease were transferred to surface work or repatriated. Compensation for accidents and industrial diseases, including miners' phthisis, was generous. In addition to the capitation fee of 6d. per calendar month per man, the WNLA advanced recruits their current tax and if necessary any arrears, paying those monies directly to the government.

Nyasaland, according to Gemmill, benefitted by at least £15 per year for every recruit. If the quota were increased to fifteen thousand, Nyasaland would receive £200,000 a year in remittances, deferred pay and cash brought back into the country. The railways and local industries would also benefit substantially.[41] The death rate had steadily decreased from 19.60 per thousand in 1937 to 5.52 in 1941. Gemmill was no doubt frustrated that, rather than gaining an increase because of the war, recruiting was suddenly suspended. After the scaling down of the military campaign in August 1943, the WNLA was granted a provisional quota of five thousand men.[42]

In July 1944 Gemmill once more visited Zomba to discuss an increase in the WNLA quota. During negotiations the Director of Medical Services raised the question of the repatriation of phthisis cases. He pointed out that, although miners received excellent compensation, no effort was made to retain them for treatment and on their return they tended to become 'foci of infection'.[43] Gemmill agreed to look into the matter and promised that if the quota were increased the WNLA might give consideration to opening sanatoria in Nyasaland. No sanatoria were established by the Chamber.

The impact

A lack of medical services meant that Nyasaland was dependent upon reports from the WNLA to monitor the health of migrant workers. An

---

[40] Gemmill, Confidential Memorandum.

[41] Gemmill, Confidential Memorandum.

[42] Précis for Executive Council, Recruitment by WNLA of Southern Province Natives, undated (1943), WNLA Correspondence LB 1/5/1, Malawi National Archives.

[43] Minutes of a Meeting of the Central Labour Advisory Board Held at Zomba on Tuesday, 25th July, 1944, M2/24/30, Malawi National Archives.

example of the way that system operated is found in the case of Henderson Abraham. In May 1942 Abraham, a labourer from the Thiramanja District, began working in Johannesburg. After nine months underground he reported sick with 'chest trouble' and was admitted to hospital. A week later he was discharged but continued to receive treatment as an outpatient.[44] When his condition deteriorated, he was repatriated, first being admitted to the government hospital in his home district of Mlanje and then being sent to the Zomba African Hospital on 10 May 1944. On admission he was diagnosed with advanced pulmonary tuberculosis and he died a few weeks later. The cost of his transport and his hospitalisation was paid by the local administration. There was no record of his having been repatriated from Johannesburg.[45] We can only guess how many times that story was repeated.

Medical care in Nyasaland was rudimentary, and only the African Hospital at Zomba had an X-ray machine and beds dedicated to tuberculosis.[46] Families tended to bring in relatives when they were severely ill, and as a result the mortality rate was very high. Although no studies had been carried out, the Director of Medical Services believed that the disease was on the increase. In 1935 there were 84 reported cases, ten years later there were 191.[47]

With the end of World War II the government wanted to conduct a survey to see whether a specialist hospital should be built. No local medical officer had the necessary expertise, and the Chief Secretary asked the Chamber of Mines if it could release a specialist to assist with the work.[48] It was a reasonable request, since the Chamber had the expertise and under the ILO convention it had some responsibility for migrant workers, but the Chamber declined. Such a survey, it said, would take two or three years and would require X-rays, clinical exams and sputum samples. It suggested that the Governor approach the Colonial Office for support.[49] In a letter of 20 October 1945 the Governor asked the Secretary

[44] Report by D. A. Baird, June, 1944, on Henderson s/o Abraham, Village Headman, Sayenda, N. A. Thiramanja District, Mlanje, Tuberculosis, M2/5/53 TB and Sanatoria Medical Report, Malawi National Archives.

[45] Letter from T. A. Austin, Director of Medical Services, to the Labour Commissioner, Zomba, 1st May 1944, Tuberculosis, M2/5/53 TB and Sanatoria Medical Report, Malawi National Archives.

[46] Letter from Dr Calleja, Acting Director of Medical Services, Zomba, to Dr Harley Williams, Secretary General, Tavistock House North, Tavistock Square, London, 29th October 1945, Tuberculosis, M2/5/53 TB and Sanatoria Medical Report, Malawi National Archives.

[47] Letter from Director of Medical Services, Zomba, Nyasaland, to Dr N. Macvicar, 46 Seventh Avenue, Parktown North, Johannesburg, 19th May, 1947, Tuberculosis, M2/5/53 TB and Sanatoria Medical Report, Malawi National Archives.

[48] Letter from D. Saunders-Jones, Acting Chief Secretary, Nyasaland Government, Zomba, to the Chairman, Chamber of Mines, Johannesburg, 7 June 1945, WNLA 59M, Medical Examinations in Nyasaland of Natives to December 1953. Teba Archive.

[49] Letter from the Acting Secretary, C. Christie Taylor, the Chamber of Mines, Johannesburg, to the Acting Chief Secretary, Nyasaland Government, Zomba, 24 July 1945, WNLA 59M Medical Examinations in Nyasaland of Native to December 1953. Teba Archive.

of State for the Colonies to fund a survey to be carried out under the auspices of the Colonial Medical Research Committee.[50] No survey was conducted.

The Chamber's refusal needs to be put in context. Speaking in London at a conference of the National Association for the Prevention of Tuberculosis in July 1947, the Secretary of State for the Colonies, Arthur Creech Jones, described tuberculosis as probably the most intractable of the diseases facing the colonies.[51] Insecticides promised to revolutionise the control of malaria, he said, but there was still no cure for tuberculosis. Colonial governments had done much to improve nutrition, housing and town planning, and they were well aware of the importance of publicity and education. Prevention, he continued, depended on a co-operative effort in which the cinema could play an important role. Positive measures were necessary to ensure that the blessing of industrial development did not bring a heavy price in disease. In the Gold Coast the government and mining interests had established a joint advisory board to deal with tuberculosis and silicosis, and he hoped that other colonies would follow their lead.

Surveys of tuberculosis had already been carried out in the West Indies and Cyprus. The government of Ceylon was conducting an anti-tuberculosis campaign, and the Fijian government had arranged for a local doctor to be given specialist training in London. The Chamber of Mines in Johannesburg, which was the major employer of migrant labour in Southern Africa, chose not to take part in such initiatives. The gold mines were at that time paying virtually no compensation to migrant workers. Between 1939 and April 1944, forty Nyasa received a total of £2,207 in compensation from the Miners' Phthisis Medical Bureau (MPMB) in Johannesburg.[52]

While it is difficult to quantify the mines' impact on public health in Nyasaland, the surviving data suggest an association between oscillating migration and a rising tuberculosis rate. In 1945 there were 332 reported cases with 38 deaths. Seventy per cent were pulmonary, and 36 of the recorded deaths occurred in that group. In that year tuberculosis was responsible for the greatest number of recorded deaths for any disease.[53] Medical officers believed that tuberculosis was on the rise. Reports during the previous twenty years had shown a threefold increase while the number of outpatients had trebled. Facilities for specialised treatment were more highly developed in the Rhodesias. Southern Rhodesia, for example, had two special hospitals with trained personnel.[54] It is signifi-

---

[50] Letter from the Governor of Nyasaland to the Secretary of State for the Colonies, 20th October, 1945, Tuberculosis, M2/5/53 TB and Sanatoria Medical Report, Malawi National Archives.

[51] Letter from the Secretary of State to the Governor, Zomba, Nyasaland, 9th July, 1947, Tuberculosis, M2/5/53 TB and Sanatoria Medical Report, Malawi National Archives.

[52] Minutes of the Central Labour Advisory Board Held at Blantyre on Friday, 14th April 1944, M2/24/30, Malawi National Archives.

[53] *Annual Report of the Medical Department for the Year Ended 31st December, 1945, Nyasaland Protectorate.* (Zomba: Government Printer, Nyasaland): 5.

[54] *Annual Report on the Public Health of the Federation of Rhodesia & Nyasaland for the Year 1955* (Salisbury: Government Printer, 1956): 11.

cant that, while the administration at Zomba was willing to export labour to Johannesburg without being able to monitor the consequences, it applied a different set of principles to white settlers. The incidence of tuberculosis among Europeans was very low, yet immigration rules required that every prospective settler and his family over the age of three years produce a radiologist's certificate of freedom from disease.[55]

Critics of oscillating migration claimed that it led to immorality and the spread of infection. The Chamber argued that, on the contrary, it protected recruits from tuberculosis and silicosis and improved their general health.[56] According to the Chamber, the WNLA and the NRC enabled migrant workers to move safely between home and the mines, explained the conditions of service (including presumably miners' rights to compensation under the Miners' Phthisis Acts), conducted medical examinations, and advanced wages and travelling expenses. The WNLA also helped with remittances and dealt with grievances. What the Chamber saw as an excellent system of medical care included monthly inspections and an exit examination. By 1947, it reported that the three hundred thousand black miners in South Africa were serviced by thirty-seven well-equipped hospitals staffed by sixty full-time medical officers. 'There can be no doubt that the tribal Native working on the gold mines of South Africa is much better off than his antitype in Asiatic countries and even many peasant communities in parts of Europe.'[57]

The opening of the new mines in the Free State in 1948 presented the Chamber with a challenge. The supply of recruits from within South Africa was declining because of competition from manufacturing industries that offered higher wages. Dissatisfaction with work contracts was also a problem.[58] Miners were paid only after the completion of thirty tickets, which took six or seven weeks. The contracts bound men to the mines for ten months, while in other industries they could give a week's notice. The mine compounds were highly regulated, which made it difficult for miners to escape their tax obligations. Recruits also resented having to repay cash advances for boots and other protective clothing.

In response the Chamber turned again to the tropical North. In 1945 10 per cent of migrant labour on the gold mines was Tropical, and by 1956 that had risen to 15 per cent.[59] The cost of transportation for local labour was £2/10/- per head as against more than £10 from Tropical areas. The industry was willing to absorb those costs because of variations in the

[55] *Annual Report on the Public Health*: 10.
[56] Simons, 'Migratory Labour' 39–40; see also *The Native Workers on the Witwatersrand Gold Mines,* Transvaal Chamber of Mines Publication P.R.D., no. 7 (1947): 17.
[57] *The Native Workers*: 5, 13–14, 19.
[58] Letter from District Superintendent to the General Manager, NRC, Johannesburg, 29 November 1952, Native Wages General, NRC 530, Teba Archives, University of Johannesburg.
[59] Memorandum, Gold Producers' Committee to the Round Table Conference on Vulnerable Mines: Recruiting and Medical Examination of Mine Native Labourers 26/9/1957, NRC 637 Miners' Phthisis Compensation to Natives 1955–1958, Teba Archives, University of Johannesburg: 2–3.

types and periods of service involved. Whereas South African labour was employed on six-month contracts and many were assigned to surface work, Tropicals worked for at least a year and always underground.[60] The high recruiting costs did, however, increase the pressure on medical officers to push recruits through the system. In January 1949 Dr R. Van Someren, the WNLA medical officer at Dowa, Nyasaland, wrote to his district manager asking for a lowering of the physical standards. There was a famine, and many of the men presenting for work were malnourished. Previously such cases had been kept back until they had gained weight, but that was no longer possible, presumably because of their very poor condition and the price of rations.[61]

In March 1951 Gemmill revisited the vexed question of who should pay travel costs. Under the ILO convention, the cost of transporting workers from Nyasaland and Barotse from Northern Rhodesia was to be paid by employers,[62] while the transport costs for other Tropical workers were paid by the worker himself. This difference was causing dissatisfaction. The gold mines had to compete with the copper mines in Tanganyika and the Wankie colliery in Southern Rhodesia, both of which paid for transport. Gemmill noted that, beyond this, the Rhodesian and Nyasaland authorities were intent on stopping the flow of labour to South Africa.

Gemmill estimated that the deductions from miners' wages for transport to and from the mines plus the costs of clothing came to around £12 a head. If the mines bore the costs of forward transport and feeding that would come to around £128,000 annually or 2s. 2d per shift. He suggested that to remain competitive the mines take on those costs.[63] It would represent a major impost, but Gemmill knew that it would allow the Chamber to minimise other costs that potentially were far more damaging. During 1956 the cost of repatriating compensated black miners with lung disease was just £13,000. A further £6,500 was spent on repatriating men who were unfit to work but had not received compensation.[64]

Cracks in the system

The return of Tropical labour from 1934 was a triumph for Gemmill and the Chamber of Mines. While governments in Pretoria supported recruitment in the North, they also inadvertently generated a body of information suggesting that the human cost was appalling. In 1943 the reports of two

[60] Memorandum, Gold Producers' Committee to the Round Table Conference: 4.

[61] Letter from R. Van Someren, Medical Officer, Dowa, Nyasaland, to the District Manager, WNLA, Dowa, 29 January 1949, WNLA 59M Medical Examinations in Nyasaland of Natives to December 1953. Teba Archive.

[62] Memo from William Gemmill, General Manager, Tropical Areas, Salisbury, to the Gold Producers' Committee, Johannesburg, 3 March 1951, Transport Costs of Tropical Natives, WNLA Repatriations and Rejects: East Coast Natives March 1925 to August 1957. Teba Archive.

[63] Memo from William Gemmill.

[64] Memorandum, Gold Producers' Committee to the Round Table Conference: 7.

important commissions of inquiry were published. The Commission on the Remuneration and Conditions of Employment of Natives on the Witwatersrand Gold Mines (Mine Wages) and the Miners' Phthisis Acts Commission or Stratford Commission arrived at the same conclusions about the gold industry. They agreed that the mines' profitability depended upon externalising the costs of production, either by paying below-subsistence wages or by not treating or compensating black miners with tuberculosis. The first strategy compounded the impact of the second on labour-sending communities.

The Commission on Mine Wages was asked to adjudicate between the Chamber's claim that wages were adequate and the case made by the Department of Native Affairs for a substantial increase on the basis of need. Gemmill represented the Chamber and as usual was a dominant voice at the hearings. He argued that in fixing wages the industry was entitled to take into account the full subsistence that migrant workers enjoyed from farming in the reserves. A migrant, he said, could earn in fourteen months sufficient to maintain himself and his family during his absence and 'to keep him in idleness for a further period of twelve months'. If necessary he could augment his income by extending his period of service. Offering higher wages would not increase the supply of labour. The reserves, Gemmill said, were practically self-supporting, and community health was good. He concluded: 'The policy [of migrant labour] is a coherent whole, and is the antithesis of the policy of assimilation and the encouragement of a black proletariat in the towns, divorced from its tribal heritage.' The gold industry was dependent on unskilled labour, and any additional working costs would shorten the lives of a number of mines.[65]

The Commission agreed with Gemmill that up to half of government revenue was derived, directly or indirectly, from the gold mines. There had been a major increase in production over the previous decade, and the tonnage of ore milled had doubled between 1930 and 1940. It also accepted that the industry was in decline and that over the coming twenty-five years many mines would close; the number of black labourers was expected to fall from three hundred and thirty-five thousand in 1945 to forty-eight thousand in 1965. It tacitly agreed that if the industry was compelled to pay a wage to support a black miner and his family living permanently in Johannesburg, then half the mines would probably close. It rejected, however, the Chamber's submissions on wages, family budgets and the cost of living, and its claims that the reserves were self-supporting and public health was good.[66]

While it conceded that in the early days migrant labour had been drawn to the mines to satisfy specific needs, the Commission argued that men were now driven by poverty. The reserves were blighted by land hunger, overstocking and soil erosion. Public health was poor, and the typical family diet was lacking in meat, milk and fats. Evidence offered by Dr

[65] *Report of the Commission on the Remuneration and Conditions*: 7.
[66] *Report of the Commission on the Remuneration and Conditions*: 2, 30, 11–14, 16.

Mary McGregor, the physician in charge of the Umtata Health Unit, was telling: nearly 50 per cent of the babies born in the Umtata District died before the age of two, mostly because of a diet saturated with carbohydrates.[67]

Subsequent research on the Victoria East/Middledrift District in the Eastern Cape for the period 1936 to 1960 supported McGregor's evidence. The overall picture of the region was one of acute poverty and overpopulation. The repeated subdivision of land meant that most households could not make a living. While there had been a population increase of 20 per cent, there had been a notable decline in the numbers aged between twenty and thirty of 15 per cent for men and 5 per cent for women. The number of older women had increased by 26 per cent with more than 30 per cent of households being headed by widows. It appeared that men were dying prematurely. Francis Wilson has offered two possible explanations: that in Xhosa society a widow always returns to her maternal residence and that a large number of men were dying from silicosis.[68]

The wages for unskilled labour offered by the railways, the Public Works Department, and the sugar industry were higher than on the gold mines. In addition, the work was less arduous and far less dangerous. Each miner was required to buy a pair of boots and a mattress for his concrete bunk and often faced additional deductions for the cost of a lamp and a protective jacket. The gold mines' ability to attract labour was due to their recruiting system and in particular the advance of £3, which enabled desperate men to provide for their families' immediate needs. The mines also offered continuity of employment. The Commission found that the average wage per shift of black miners had been higher in 1925 than it was in 1942 and that wages had fallen well below the subsistence needs of men and their families. It recommended a flat rate wage increase for all black workers of three pence a shift.[69] There was no rise.

As we have seen the Stratford Commission report of 1943 presented a damning assessment of the mines and their role in spreading tuberculosis. It focused public attention on the migrant labour system, and there were allegations in the press that black miners were spreading tuberculosis. In response, in May 1944 the GPC commissioned a series of newspaper articles. Each item was written by a Chamber official, Mr H. Husted, for a particular newspaper and with a particular target audience in mind.[70] The articles shared a number of themes: that the tuberculosis rate among miners was falling, that disease was widespread in rural areas because of poor hygiene and government inaction and that urbanisation was a greater

---

[67] *Report of the Commission on the Remuneration and Conditions*: 14.
[68] Wilson, *Migrant Labour*: 102.
[69] *Report of the Commission on the Remuneration and Conditions*: 20.
[70] Articles were written for the *Star*, the *Sunday Times* and the *Forum*. Memo from A. Limebeer, Secretary of the GPC, to Members of the Gold Producers' Committee, subject Health in Mine Workers 4 May 1944 Miners' Phthisis Compensation for Natives 1942–1949 NRC 390 1 & 2, Teba Archives, University of Johannesburg.

threat to public health than the mines. The piece for the *Rand Daily Mail* took the form of an interview with an executive member of the MMOA. He was quoted at length as saying that tuberculosis was endemic in the Native territories and that the mines played no part in the spread of disease: 'In the end, of course, it is the malnutrition and unhygienic living conditions in the Territories that should be fought.' The article contained no attribution to the Chamber, and readers were not told that the interview was fictitious.

In February 1944 Peter Allan, by then Secretary for Public Health, wrote to the Chamber that repatriated miners were spreading tuberculosis in the Eastern Cape. His department, he said, was so concerned that it was building a specialist hospital at Umtata and had set up a reporting scheme with the Durban health authorities. 'The problem is obviously very serious,' he wrote, 'and the number of tuberculous natives returning to the territories may seriously influence the position there. The solution is very difficult as any form of isolation would have to be enforced. There are obvious objections to this including the fact that it would presumably be detrimental to recruiting.' He asked the Chamber for financial help.[71] In response the GPC set up a special subcommittee whose members included Girdwood of the WNLA and the Chamber's legal adviser. In the absence of any new research, discussion fell back upon the official disease rate. During the previous three years the rate of repatriation for tuberculosis had been 0.24 per cent, 0.28 per cent and 0.33 per cent, respectively. On the basis of that evidence the subcommittee concluded that the mines were not spreading tuberculosis. 'It is well known,' its minutes state, 'that the Natives are heavily tubercularised and that any change from the normal leisurely open air life of the kraal may cause a breakdown in their physical condition leading to active tuberculosis.' The reasons for infection were ignorance, malnutrition and a lack of medical services in rural areas, all of which were the responsibility of government. Infected miners wanted to return home, and it was impossible to persuade them to remain in treatment. 'In the view of the Committee, the solution lies in the education and enlightenment of the Native people themselves.' That could be assisted by the establishment of government-funded health clinics. There were in addition other serious health problems, such as venereal disease, ravaging native communities which demanded government action.[72]

Outside South Africa, there was also criticism of the mines. The daily papers in Lourenço Marques featured a number of articles accusing the gold mines of spreading tuberculosis in Mozambique. Informed of those

[71] Letter from Dr Peter Allan, Secretary for Public Health, to the Secretary, Chamber of Mines, Johannesburg, 1 February 1944, WNLA 20L Tuberculosis February 1931 to February 1957 Diseases and Epidemics. Teba Archive.
[72] Minutes of the Special Sub-Committee of the Gold Producers' Committee to Consider 'Repatriation of Native Tuberculotics from the Mines to the Native Territories' 7th and 14th March 1944, WNLA 20L Diseases and Epidemics, Tuberculosis February 1931 to February 1957. Teba Archive.

reports, the WNLA's General Manager, G. O. Lovett, was unsympathetic. Tuberculosis was endemic, he said, and infection on the mines had been constantly declining, the current rate being a mere 0.21 per cent. There was no evidence that the mines were spreading disease. Breaks in service protected migrant labour against miners' phthisis and tuberculosis. Tuberculosis among blacks was rising at an alarming rate in urban areas. It was a disease of squalid, overcrowded urban communities and poor diet, not a disease of the mines.[73]

The problem of lung disease in miners was examined by the commissions of inquiry into silicosis and tuberculosis conducted in Southern Rhodesia in 1937 and Northern Rhodesia in 1949. Both acknowledged the synergy with silicosis and made recommendations on how to contain the spread of infection. Southern Rhodesia had a large gold mining industry, and in 1937 the Medical Director, Dr A. P. Martin, completed a preliminary report on silicosis and tuberculosis in which he concluded: 'To the mine native the greater danger in mine work is the danger of contracting tuberculosis during the course of his employment, either directly or in association with some amount of silicotic fibrosis. One must not forget also the probability that a majority of the cases originally classified as silicosis who may leave mining may eventually develop tuberculosis.'[74]

The Northern Rhodesia Commission of 1949 reached the same conclusions. African workers with tuberculosis were often sent to the government hospital at Ndola. In some cases they would stay only a few days before returning home. From that moment their diet would deteriorate and they would have no medical supervision. Some miners would arrive home with open tuberculosis, a menace to their families and communities. The Commission recommended that the state urgently respond to tuberculosis as a public health problem, that examinations be carried out annually and that they include both radiographic and clinical reviews. Every African leaving the mines should be fully examined by the Bureau: 'We are satisfied that full-scale radiographic examination, if financial and other practical circumstances permit its arrangement, is in every way to be preferred to the use of a miniature plant which cannot, for example, be relied upon to show the presence of silicosis in the very early stage.'[75]

---

[73] Letter from G. O. Lovett, General Manager, WNLA, to the District Manager, WNLA Lourenço Marques, 5 November 1945, WNLA 20L Tuberculosis February 1931 to February 1957 Diseases and Epidemics.

[74] *Report of the Commission Appointed to Enquire into the Possible Prevalence and Origin of Cases of Silicosis and Other Industrial Pneumoconioses in the Industries of the Colony of Southern Rhodesia and of Pulmonary Tuberculosis in Such Industries* (Salisbury: Government Printer, 1938): 16.

[75] *Report of the Commission on Silicosis Legislation* (Lusaka: Government Printer, 1949): 68, 10, 11–12.

Post-mortems

The mines controlled the system of medical surveillance and data collection on which their reputation for safety was based. According to the official returns during 1928, there were 862 cases of tuberculosis among black miners, an incidence of 4.50 per 1,000. By 1942 the incidence had fallen to 2.54.[76] The Annual Report of MPMB for 1941 noted that the silicosis rate had been constant over the previous ten years. The rate of approximately 2 per 1,000 among black labourers was, the Bureau admitted, not comparable with the incidence of 10.05 per 1,000 for white miners: whites worked permanently underground and therefore had higher exposures to silica. To an outsider the discrepancy between the two sets of data is remarkable.[77] Whites had lower dust exposures and their physical requirements for mine employment were also lower.

The achievements of South Africa's medical system were on show in London in April 1947 where a conference on silicosis, pneumoconiosis and dust suppression was sponsored by the Institute of Mining Engineers, and South African specialists gave five of the seventeen papers. Dr J. M. Smith, chair of the Silicosis Medical Bureau (formerly the MPMB), spoke about the Bureau's achievements before an audience drawn from the UK, the USA, Australia, Belgium, France, Germany and Norway. Smith's data showed an inexorable decline in silicosis and tuberculosis among white miners: in 1927 the silicosis rate was 1.9 per cent; in 1944 it had fallen to 0.8 per cent. Smith told his audience that owing to a lack of time his account of black miners was necessarily brief. For black miners the figures in 1944 were 0.28 per cent for tuberculosis and 0.24 per cent for silicosis.[78]

The official data acted as a barrier to criticism of oscillating migration and work conditions underground. There was, however, one notable crack in that edifice. Under the Miners' Phthisis Acts Consolidation Act of 1925, any miner who died suddenly was subject to a post-mortem. Between 1925 and 1950 the results were published in the annual reports of the MPMB. During that period around six hundred miners perished each year in accidents, and presumably most of the autopsies were performed on that random group. The data from those autopsies show far higher rates of silicosis and tuberculosis than were identified in living miners. That in itself is not surprising, as it had long been acknowledged that post-

---

[76] *Report of the Miners' Phthisis Medical Bureau for the Three Years Ending 31st July 1941*, UG, No. 18 (Pretoria: Government Printer, 1944): 28.

[77] *The Prevention of Silicosis*: 224.

[78] J. M. Smith, 'Review of the Work of the Silicosis Medical Bureau, Johannesburg' in *Silicosis, Pneumoconiosis, and Dust Suppression in Mines: Proceedings of the Conference Held in London April 1947* (London: Institute of Mining Engineers, 1947): 92, 95.

mortems might uncover lung disease that had been missed in clinical or X-ray examinations. What are surprising are the dramatic differences between the two sets of data.

The data sequence from 1924 until 1950 revealed a tide of disease. In 1924 the lungs of 122 white and 176 black miners were examined. Silicotic changes were found in 97, or 79.5 per cent, of whites. Among that group 28 had been certified as free of silicosis, some within months of death. The lungs of the 176 blacks showed an even higher rate: in 78 the cause of death was tuberculosis even though the deceased had been subject to periodic medical examinations. In another 60 cases tuberculosis with or without silicosis was present.[79] In 1928 the lungs of 227 white miners and 429 blacks were examined. Of the whites 55 per cent had lung disease, while 81 per cent of the blacks were affected.[80]

In 1941 post-mortems were performed on 999 black miners. Of that cohort only 162 were free of silicosis or tuberculosis. In total, 170 had both diseases while 606 had tuberculosis alone. At death 84 per cent of the cohort had a compensatible disease.[81] The context is also significant. Between 1931 and 1941 there was a sharp rise in the number of black miners. Consequently, many of the deceased who appear in the 1941 report were probably new to the industry. Of the 915 lungs examined in 1949, 327 were from whites and 584 from black labourers. Of the white miners 43 per cent had some form of disease, a marked increase that coincided with the war years. Of the black miners 83 per cent were suffering from silicosis and/or tuberculosis.[82] In the following year, of the 649 blacks examined 370 or 55 per cent had tuberculosis without silicosis and the total disease rate was 82 per cent.[83]

The Bureau's annual reports made little comment on the post-mortem results. In his yearly summary the Director usually focused on the improvements in mine safety, the low rates of silicosis and tuberculosis, the enhanced medical surveillance, the lessening severity of silicosis, the research initiatives and the protection afforded blacks by their short-term contracts. After more than twenty years of post-mortems, in 1950 the Bureau resolved the problem by deleting the results from its Annual Report. In place of the data the Reports for 1952 and 1953 discussed at length the factors that made it difficult to arrive at an accurate disease rate. The Director explained that the daily average number employed in dusty work tended to be unreliable, the service distribution of labour was unknown and the periodic examinations did not always pick up early

[79] *Report of the Miners' Phthisis Medical Bureau for the Twelve Months Ending July 31, 1924* (Pretoria: Government Printer, 1925): 4.
[80] *Report of the Miners' Phthisis Medical Bureau for the Year Ended the 31st of July, 1928* (Pretoria: Government Printer, 1929): 7.
[81] *Report of the Miners' Phthisis Medical Bureau, July 1941*: 7.
[82] *Report of the Silicosis Medical Bureau for the Year Ended 31st March 1949* (Pretoria: Government Printer, 1950): 6.
[83] *Report of the Silicosis Medical Bureau for the Year Ending 31st March 1950* (Pretoria: Government Printer, 1951): 8.

signs of silicosis.[84] Of course, none of those factors explained the post-mortem results.

The post-mortem results from 1924 to 1950 are consistent with the science produced since majority rule in 1994 in that both identify a disease rate at least one hundred times higher than the official rate. They suggest that the exit medical examinations were inadequate, the compensation system was flawed and tuberculosis was being spread from the mines. For decades those data were known to the MPMB, the MMOA, the Department of Mines and Health, the Chamber of Mines and the specialists at the SAIMR.

---

[84] See *Report of the Silicosis Medical Bureau for the Two Years Ended 31st March 1952 and 31st March 1953* (Pretoria: Government Printer, 1955).

7

Conflict over the Compensation System

The opening of the new gold fields in the Free State was the largest single investment in the industry's history. Between 1946 and 1950 Anglo American poured more than £65,000,000 into the development of thirteen new mines.[1] It was a massive undertaking designed to ensure the corporation's future and it coincided both with the advent of apartheid, which extended and deepened existing inequalities, and with a period of intense conflict between the white trade unions and industry. That conflict began with the passage of the Silicosis Act No. 47 of 1946 and ended twelve years later with the Pneumoconiosis Act No. 57 of 1956.

The Mine Workers Union (MWU) believed that the Silicosis Medical Bureau had a policy of deliberately minimising the number of awards. In 1951 it complained in the *Rand Daily Mail*: 'Not one per cent of mine employees would pass a vote of confidence in the Bureau which has today become a white elephant in every sense of the word.'[2] The underlying problem was the intractability of dust and the refusal of the Bureau to compensate white miners with what the MWU considered palpable disease. Under apartheid, black miners lay largely outside of that struggle.

On the surface the system of medical review and compensation appeared to be working smoothly. The industry-generated data for the period 1945 to 1955 show that the tuberculosis rate among long-term black miners, those with poor physiques and new recruits X-rayed at the Witwatersrand Native Labour Association (WNLA) depot was 0.1 per cent. The silicosis rate had fallen from 1.3 per cent to 0.5 per cent.[3] Of the 290,183 miners X-rayed at the WNLA depot in 1956 only 3,976 cases were referred

---

[1] 'Chairman's Report' in *Annual Report of Anglo American for the Year Ended 31 December 1950*: 2.
[2] See *Report of the Medical Bureau for Occupational Diseases for the period 1 April 1975 to 31 March 1976*, South African National Archives: 1.
[3] Annexure C, WNLA Miniature Radiographical Survey of Natives with Long Labour Histories or Poor Physiques and New Recruits at WNLA Depot, in Memorandum: Gold Producers' Committee to the Round Table Conference.

to the Bureau for review.[4] Despite the promising data, there were tensions between the Bureau and industry. In February 1949 Dr S. W. Verster, chairman of the Bureau, met with the Group Medical Officers to discuss the entry and exit medicals. Verster commented that *ex gratia* examinations by district surgeons often revealed silicosis or tuberculosis in men with long underground service. 'Evidently there seem to be cases that have not been X-rayed when repatriated, as quite often these are detected within a year or eighteen months after leaving the mines.' He also complained of the inadequate entry medicals which saw a number of miners being repatriated with tuberculosis after working as few as six days. Some of those men were fresh recruits, while others had had previous employment underground.[5] Whatever their backgrounds, they had been ill when recruited. The Group Medical Officers agreed that more care was needed in conducting medicals and that X-rays should be used more widely.

The introduction of mass radiography after 1950 extended the reach of what Orenstein had earlier described as 'mass production medicine'. The technique enabled a miniature X-ray to be taken cheaply and quickly. It was used initially with selected groups such as black miners with long service and then gradually extended to all migrant workers. On average a medical officer at the WNLA would read between eight hundred and a thousand X-ray plates in an hour. Dr J. A. Louw proudly told a Mine Medical Officers' Association meeting in 1965 that in an hour he could review up to two thousand films.[6] His achievement is reminiscent of the Woody Allen joke about his having taken a speed-reading course that allowed him to read *War and Peace* in a sitting. When asked what it was about, he replied, 'Russia'.

The Chamber of Mines took care to present a positive image to the outside world. In July 1952 the WNLA's Chief Medical Officer, Dr Frank Retief, speaking at the Third Commonwealth Health and Tuberculosis Conference in London, gave a glowing report on the WNLA's tuberculosis programme and outlined the advantages of miniature radiography. In 1948 the mortality rate for pulmonary tuberculosis among miners was 0.5 per cent, a figure that, as Retief remarked, compared favourably with the data for the UK population as a whole. The South African system of medical surveillance had no parallel, he said, and when miners became ill they were liberally compensated. Each year more than fifty-six thousand recruits were X-rayed, and the programme was being expanded to include all new entrants from north of 22 degrees South latitude. The new equipment could X-ray eight hundred men an hour, and eventually three

[4] Memorandum: Gold Producers' Committee to the Round Table Conference: 6.
[5] Minutes of Meeting of the Sub-Committee of the Group Medical Officers 1 February 1949, topic: Mines Native Labourers: Diagnosis of Tuberculosis, NRC 390 1 & 2 Miners' Phthisis Compensation for Natives 1942–1949, Teba Archives, University of Johannesburg: 3.
[6] Dr J. A. Louw, 'The Miniature Chest Radiograph: Its Uses, Value and Problems' *Proceedings of the Mine Medical Officers' Association* 45, no. 392 (1965): 81.

hundred thousand exams would be done annually. Retief told his audience: 'We consider that the gold mining industry of the Witwatersrand which employs large numbers of persons is the only industry of its size in which all persons seeking employment, whether previously employed in the industry, or not, will be automatically X-rayed.'[7] Retief repeated the same story during an interview on BBC Radio, with one difference. Much to his annoyance the BBC insisted that he substitute 'Africans' for 'Natives' throughout his talk.[8]

The passage of the Pulmonary Disability Act in 1952 saw the number of initial certifications for white miners rise from 306 in 1953 to 746 in 1954.[9] The industry was concerned that the trend would continue, and in February 1954 the Chamber of Mines' President, K. Richardson, and his colleagues B. L. Bernstein and H. Husted met with the Minister of Finance, N. C. Havenga. Richardson explained that the Chamber had grave doubts about the competence of the Pulmonary Disability Committee. When the Committee was established the Chamber had expected the extra cost to be around £400,000 to cover past cases and an additional £400,000 a year for new certifications.[10] According to the Chamber's estimates, those figures could treble, meaning a rise of approximately 8d. per ton in operating costs spread over the industry as a whole. The older mines could not withstand such an increase and some would be forced to close. The position, he emphasised, was grave. The Minister was sympathetic and agreed that the matter would probably have to be discussed in Cabinet. Two hours later Richardson and his colleagues met with the Prime Minister, D. F. Malan, with whom he raised the same issues. During a cordial discussion Malan said that he took a special interest in silicosis and pulmonary disability compensation and 'ended the interview with the observation that the gold mining industry was undoubtedly the economic backbone of the country'.[11]

Within weeks an inquiry was convened under the chairmanship of Prof. S. F. Oosthuizen of the University of Pretoria to resolve what the

[7] 'Miniature Radiography of the New Native Entrants to the Gold Mining Industry on the Witwatersrand' paper presented by Dr Frank Retief, Chief Medical Officer, WNLA, to the Third Commonwealth Health and Tuberculosis Conference, London 8 July 1952, WNLA 20L Diseases and Epidemics, Tuberculosis February 1931 to February 1957, Teba Archives, University of Johannesburg: 1, 3.

[8] Transcript, 'Medical Activities of the Gold Mining Industry in South Africa' Dr Frank Retief 10 July 1952, BBC London, WNLA 20L Diseases and Epidemics, Tuberculosis February 1931 to February 1957, Teba Archives, University of Johannesburg.

[9] It rose again to 1,466 in 1955. With the repeal of the act the numbers fell to 399 in 1958 and 186 in 1959. See Table 3 in *Report of the Pneumoconiosis Compensation Commissioner for the Year Ending 31st March 1959* (Pretoria: Government Printer, 1959): 7.

[10] Minutes of Meeting with the Hon. Minister of Finance, Mr N. C. Havenga, Wednesday, 17 February, 1954, at 9:30 a.m. WNLA 20L Diseases and Epidemics, Tuberculosis March 1953–Nov. 1954, Teba Archives, University of Johannesburg: 1.

[11] Meeting with the Hon. Prime Minister, Dr D. F. Malan, Wednesday, 17th February, 1954, at 11:30 a.m. WNLA 20L Diseases and Epidemics, Tuberculosis March 1953–Nov. 1954, Teba Archives, University of Johannesburg.

Chamber viewed as a crisis. Although evidence was heard in camera, it was not taken under oath. Oosthuizen was asked to inquire into the relationship between silicosis and pulmonary disability as defined under the Silicosis Act, to determine whether it was desirable to introduce a single comprehensive definition of disability, and, if so, to draft such a definition and to recommend measures for effective diagnosis.[12] The narrow terms of reference were reflected in the witnesses, most of whom were radiologists, clinicians, pathologists and dust experts. The MWU and the mining union's joint committee declined to appear. Oosthuizen had no brief to investigate the functioning of the Silicosis Medical Bureau or the Pulmonary Disability Committee. The Chamber's Group Medical Officers' subcommittee formulated a strategy to be used at the hearings, and once Oosthuizen's report was released the Gold Producers Committee (GPC)'s legal secretary, D. R. Worrall, wrote a review of the policy implications.[13]

The Oosthuizen inquiry spent much of its time investigating complaints about the compensation system. The most serious allegations concerned the poor quality of the Silicosis Bureau's medical examinations and the role of the Pulmonary Disability Committee under the directorship of Dr G. W. H. Schepers. The inquiry determined that the Bureau was failing to accept legitimate claims while the Pulmonary Disability Committee was accepting claims that had little merit. In addition to its formal deliberations, the inquiry submitted a confidential report to the Minister of Mines, Dr A. J. R. van Rhijn, accusing Schepers of seeking to gain control of the Bureau and of having dragged 'the good name of the South African medical profession in the mud'. The report described Schepers as a man whose outstanding intellect was compromised by an 'inexcusable scientific dishonesty'. As Director of the Committee, it said, Schepers had been in conflict with the Bureau, the Silicosis Medical Board of Appeal, the Silicosis Board, the Department of Mines and the South African Institute of Medical Research (SAIMR), and had undermined the authority of the Bureau, the Secretary for Mines and the Minister. In addition, it went on, he had engaged in unethical behaviour in having leaked confidential information to the MWU. The report warned that Schepers' continued employment would lead to serious industrial unrest.[14]

Oosthuizen was a powerful figure in the medical establishment. In addition to holding a chair in radiology at the University of Pretoria, he was at that time President of the South African Medical and Dental

[12] *Report of the Departmental Committee of Enquiry into the Relationship between Silicosis and Pulmonary Disability and the Relationship between Pneumoconiosis and Tuberculosis* (Pretoria: Government Printer, 1954): 3.

[13] See Memo: Silicosis Act: Department Committee Tuberculosis, from D. R. Worrall, Assistant Legal Secretary, to Members of the Gold Producers' Committee, 22 April 1955, NRC 6066 Miners Phthisis' Compensation for Natives 1954–55, Teba Archives, University of Johannesburg.

[14] *Supplementary Confidential Report of the Departmental Committee of Enquiry into the Relation tion Between Silicosis and Pulmonary Disability, Departmental Committee to Inquire into the Definition of Silicosis & Chest Diseases (Oosthuizen) Departmental Committee 1954*, South African National Archives, Pretoria, F 33\671: 14, 23, 35, 30, 40.

Council and Secretary of the Medical Research Council. He and Schepers knew each other well. Schepers had appointed him to the Pretoria Medical School. Oosthuizen had taught Schepers' wife, and he frequently visited the Schepers household in Johannesburg. After Schepers and his family emigrated to the USA he invited Oosthuizen to attend a conference in Saranac, in upstate New York. During that visit Oosthuizen asked Schepers to help him find a job in the USA.[15] Schepers only became aware of Oosthuizen's attack on his character when I sent him a copy of the confidential report in September 2008.

Schepers had left South Africa in 1954 to take up the directorship of the Saranac Laboratories, the leading pulmonary research centre in North America. The appointment was a major career advance and during more than four decades in the USA Schepers enjoyed uninterrupted success as a researcher and administrator. If he had remained in Johannesburg he might well have suffered much the same fate as Dr George Slade, whose pioneering research on asbestos disease in 1930 destroyed his career.[16]

The silicosis medical bureau in the 1940s and 1950s

Schepers had received a first-class education at the University of the Witwatersrand, where he studied under the pathologist Sutherland Strachan.[17] For a time after receiving his doctorate in medicine in 1948, he was professor of pathology at the University of Pretoria, but he had no particular wish to specialise in pulmonary medicine. Because of a leaking heart valve he was disqualified from military service in World War II and was assigned in 1944 to the then Miners' Phthisis Medical Bureau in Johannesburg.[18]

Initially the Bureau had not been subject to civil service regulations and therefore had some genuine independence from the Department of Mines and the Chamber. The Silicosis Act of 1946 brought the Bureau under the control of the Department and made its chairman directly responsible to the Secretary of Mines. That restructuring eroded the Bureau's independence at a time when the MWU was lobbying the state for increased compensation. The Bureau's main function of performing medical examinations of white miners remained unchanged. Its doctors took turns each week examining miners at the WNLA hospital who were suspected of having silicosis or tuberculosis. It was seen as an onerous task by the interns, but Schepers was interested in the work and spoke three indigenous languages, so he volunteered. He hoped to improve the

---

[15] Interviews with Dr Gerrit Schepers, Great Falls, Virginia, 23–28 October 2010.

[16] See McCulloch, *Asbestos Blues*: 120–22.

[17] This account of Schepers' career is based in part on his deposition at Alexandria, the US District Court for the Eastern District of Virginia, 29 June 1981 in re: Asbestos Litigation Locke v. Johns-Manville C/P No. 77–1.

[18] It was an inaccurate medical review. Schepers lived on into his nineties in excellent health.

certification system for, as he later recalled, 'I cannot remember a single diagnosis for silicosis at the WNLA until I did the work.'[19] ('Internship' was the term used at that time to describe the status of the Bureau doctors.)

White miners were processed at the Bureau in groups of around three hundred a day. The workloads were crushing, and in 1951 the Bureau's ten interns each carried out 5,472 examinations.[20] Allowing for other duties including a daily team meeting, that gave interns at most twelve minutes to examine each man, read his X-rays and make a judgment about his health. Interns were obliged by law to complete a form containing a full medical and work history, but because of understaffing such details were not always recorded.[21] Schepers found the Bureau doctors 'careless and unfeeling' and was astonished at the way medical reviews were conducted. There were ten interns, and in the mornings they examined white miners. In the afternoon there was a meeting with the Director as chair. The interns sat at a round table, and the X-rays for each applicant were circulated on a rotating screen. Compensation decisions were made solely on the basis of the number and size of nodules in a miner's lungs. There were hundreds of plates to be read, and each was given only a few seconds. The Bureau's machines had a broad focal point, which was adequate for identifying large silicotic densities but could not pick up the small nodules characteristic of early-stage silicosis. Consequently, only men with massive fibrosis were given compensation.

Under the law, Grade Two silicosis was based on disability, but there was no lung- function equipment and therefore no way to measure lung capacity. The budget at the Bureau was limited, and there was no money for pathology testing. Each doctor voted, with the Chairman having a casting vote when necessary. Schepers routinely voted in favour of compensation, which meant that he voted against the majority view. He believes that in 90 per cent of cases the Bureau's decisions were probably wrong. During his years as an intern the official silicosis rate for white miners was around 0.2 per cent. He suspects that the actual rate was a hundred times higher: 'The system of medical review was organised fraud.'

Schepers wanted to see the introduction of a system based on 'the demonstrable deterioration of cardio-pulmonary function' rather than an abstraction such as the number of nodules visible on an X-ray plate. His criticism of the Bureau's methods brought him into conflict with Dr Verster.[22] In recalling that period Schepers says that he had no interest in politics and his opposition to the system was based on medical ethics not

[19] Interviews with Dr Gerrit Schepers, Great Falls, Virginia, 23–28 October 2010.
[20] *Report of the Commission of Enquiry into the Functioning of the Silicosis Medical Bureau and the Silicosis Medical Board of Appeal* (Pretoria: Government Printer, 1952): 7.
[21] Unless otherwise indicated, the following account comes from interviews with Dr Gerrit Schepers.
[22] Dr Gerrit Schepers, transcript of evidence before the Beyers Commission, 29 & 30 January 1952, Teba Archives, University of Johannesburg: 505–6.

ideology. 'I was not a radical,' he recalls. 'I was not political. I only wanted to practice medicine properly. My family was Christian, and it was those values which put me at odds with the authorities.'

During Schepers' years at the Bureau it was common for white miners with serious lung disease to be refused compensation. What happened at the WNLA was far worse. In theory the men were to be examined by mine medical officers who would refer suspected cases to the Bureau for adjudication. The X-rays were read, however, by the WNLA doctors, not by the Bureau interns. Medical officers had no indigenous languages and were unable to ask migrant workers even simple questions about their symptoms and work experience. Black miners did not attend the Bureau, and the decisions on their right to compensation were made at the WNLA compound.

The X-ray equipment at the WNLA was adequate, but there was a massive volume of work. The miners were lined up and a doctor would run a stethoscope over each chest, doing ten men in a minute. They were not given a clinical examination, and only those with obvious disease were X-rayed. The other test was weight loss. The exit examinations were conducted at the mines. In 1950 that was the beginning and end of medical surveillance.

Schepers recalls that there was a lot of disease and in many cases the black miners he saw at the WNLA were 'in the process of dying'. Some of the men had simply been worked to death. After two years underground a man's health was usually ruined: 'Whites survived on average three years after retirement. Blacks who had worked three consecutive contracts were often dead a year after they left the mines.' The rate of tuberculosis was very high, but most cases were repatriated without compensation. According to Schepers 'the horrendous spread of tuberculosis from the mines' was common knowledge at the Bureau.

There was no free exchange of opinions among the interns, and no minutes were kept of meetings. There was no follow-up of individual cases. While the Chamber wanted to minimise the cost of compensation, it did not exert any direct influence over the Bureau's daily operations. Instead, according to Schepers, Orenstein at the GPC 'ran the Bureau's head doctors' through the office of the Minister of Mines. The Bureau staff members were required to sign a confidentiality agreement and faced a £10,000 fine or ten years' imprisonment if they talked to the press. As most of the interns were British and close to retirement, they were reluctant to risk their pensions by opposing the Chairman.

The Silicosis Act required that the Bureau certify silicosis where there was 'the least evidence of disease'. The issue was not fitness to work but the presence of silicosis. In practice the Bureau used the capacity to work as the criterion for keeping men underground as long as possible before they were pensioned off. Schepers recalls: 'The attitude of the Bureau was that certification should wait until a man was no longer capable of working. Wait until he is sick and can't work.' To receive compensation a miner would need a major lesion and massive nodules on his lungs.

In 1949 Schepers received a fellowship from the Harkness Foundation to study in the USA. He enrolled at the Institute of Industrial Medicine at New York University, where his supervisor was Dr Anthony Lanza, the leading authority on silicosis and asbestos disease. On his return to South Africa in September 1950 Schepers filed a report on what he had learned with Dr Verster the chairman of the Bureau and with V. H. Osborn, the Secretary of Mines. They showed no interest. Schepers pressed for reform of the compensation system and in particular for the inclusion of emphysema and bronchitis under the award. He felt strongly that the newly-elected government of the time was a tool of the mining houses and that the Bureau was engaged in covering up occupational disease. Through the Johannesburg attorney, Albert Hertzog, he met Daan Ellis, the Secretary of the MWU. Ellis had been dismissed from his job as a schoolteacher following a scandal. He became a miner and rose quickly in the union leadership. Ellis was helped by his powerful connections in the National Party.

Ellis and Schepers began lobbying for change. Schepers introduced a former Bureau intern, Dr Inglethorpe, to Ellis, who employed him as the MWU's physician. On Schepers' advice, the union bought a costly X-ray machine to carry out its own medical assessments. After work Schepers regularly went to the MWU and reviewed the files of miners who had been refused compensation. Schepers and Inglethorpe soon built up a series of a hundred cases based on X-rays and tissue samples of miners whom the Bureau had wrongly found to be free of silicosis or tuberculosis. Schepers also obtained tissue specimens from black miners who had died at the WNLA hospital. The mine doctors were required by law to send the lungs of deceased miners to the WNLA, but because the latter had no pathology service the tissues were usually discarded.

In 1952 Schepers wanted to appear before the Beyers Commission, which was investigating the functioning of the Silicosis Medical Bureau, but was forbidden to do so by the Bureau's chairman and then by the Minister of Mines. In frustration, he lobbied members of Parliament and at one point met with the Prime Minister, D. F. Malan. Eventually he did testify highlighting a number of changes in the Bureau's procedures which he viewed as obstructive.[23] Prior to May 1951 the lungs of deceased white miners had been sent to the SAIMR, where experienced pathologists determined whether silicosis was present and whether it had contributed to the man's death. Its reports and tissue samples were forwarded to the Bureau for final assessment. Each afternoon the interns would examine three or four pairs of lungs and come to a decision as to the cause of death.[24] On 1 May 1951 Verster met with the SAIMR pathologists Murray and Becker and informed them that, to avoid influencing the Bureau's

---

[23] See Schepers, transcript of evidence before the Beyers Commission, 505–6, 511. In the author's possession.

[24] Schepers, transcript of evidence before the Beyers Commission: 529.

decisions, in future they should express no opinion on the presence of silicosis or tuberculosis and make no comment on bronchitis or emphysema.[25] It was a significant change, because the pathologists had expertise that the Bureau lacked. There had been a problem with the Institute's pathologists finding silicosis and the Bureau's deciding that it was not present. After 1 May, Verster alone made those assessments, and they were recorded as Bureau decisions. When Schepers protested to the Secretary of Mines, he was told that the Secretary had authorised the initiative.[26] Verster's new policy gave the Bureau *carte blanche* in its determinations.[27] There was no longer an external body of review.

At the same time as the post-mortem system was changed, the Bureau Chairman assumed the right to both a casting and a deliberative vote on compensation cases. Where a decision was disputed, the Chairman would use his two votes to determine the outcome. Verster already controlled the leave privileges and the promotion prospects of the interns, and those changes increased his influence. According to Schepers, from that point Verster had so much control over certifications that members of the Bureau sometimes voted against their own consciences.[28]

In its final report the Beyers Commission agreed with Schepers that the Bureau's medical reviews did not conform to the Silicosis Act, largely because of crushing workloads and understaffing. It also questioned the merits of X-rays as a means for assessing claims. Despite objections from two of its members Drs Orenstein and Oosthuizen, the Commission supported the introduction of emphysema and bronchitis as compensatible diseases. At the request of the Deputy Prime Minister, Schepers wrote the original draft of the Silicosis Amendment Act (Pulmonary Disability Act) No. 63 of 1952, which was passed while the Beyers Commission was still in session.

Under the new legislation, Schepers was appointed Director of the Pulmonary Disability Committee, a newly created arm of the Bureau. He assembled a team consisting of a cardiologist, Dr Ian MacGregor, a pulmonary physiologist, Dr Margaret Becklake, a biostatistician, and two internists, Drs Sluis-Cremer and Duthie. The Committee developed a more rounded approach based on lung function testing and electro-cardiographic deficits. The result was a dramatic rise in the number of successful claims and a breakdown of relations between Schepers and Verster.

In February 1954, Schepers gave evidence to the Oosthuizen Committee. At the centre of his testimony were the cases of four men – Cronje, C. J. Pretorius, Coetzer and Paens – who had applied for compensation and, despite evidence of disease, had been rejected. The case of Pretorius, who died of a pulmonary haemorrhage at the age of forty-three,

[25] Schepers, transcript of evidence before the Beyers Commission: 531.
[26] Schepers, transcript of evidence before the Beyers Commission: 531.
[27] Schepers, transcript of evidence before the Beyers Commission: 530.
[28] Dr G. W. H. Schepers, transcript of evidence before the Oosthuizen Committee, Johannesburg, 22 February 1954, Teba Archives, University of Johannesburg: 972.

caused serious conflict between Schepers and Verster.[29] The Cronje case was equally disturbing. Cronje had a history of progressive breathlessness, a chronic productive cough and extreme lassitude. He was examined at the Bureau in June 1953, but despite his symptoms, which were suggestive of tuberculosis, no sputum sample was taken. The Bureau's interns also found Cronje free of silicosis. In August Schepers examined Cronje and reviewed his X-rays, which had been taken over an extended period. He found a lesion on Cronje's left lung that went back some years.[30] Schepers asked the Bureau radiologists to re-examine Cronje, but they refused.

Schepers was asked by the Oosthuizen Committee if he held any one person responsible for the system's failings. He replied:

> During the period from 1944 to about 1947, I myself observed a progressive deterioration in the – how shall one put it as kindly as possible – preparedness of the Bureau to admit diseases when they are present. This is reflected, actually, in the statistics of the Bureau. You will find that if you analyse our Bureau reports, that during that phase more and more tuberculotics were first certified after they were dead, than during life.[31]

Fifteen years earlier, 40 per cent of tuberculotics had been certified on the basis of disability, but by 1948 that had fallen to fewer than 5 per cent. At the same time the number of cases identified at post-mortem had risen sharply.[32] Schepers went on to comment: 'Certain members of the Bureau displayed a really consistent reluctance to admit the diagnosis of either silicosis or tuberculosis in too many cases. So that I was invited more or less to infer that their reluctance to make those diagnoses was a manifestation of policy, rather than ignorance or inefficiency.'[33] He believed that the actual number of cases was rising. When asked about the Bureau's motives Schepers replied: 'One got the impression that it was for the benefit of the industry – not losing their workmen and not disbursing compensation unduly generously.'[34]

The Oosthuizen report had an immediate impact. The Pulmonary Disability Committee was shut down and the Pulmonary Disability Act rescinded. Its replacement, the Pneumoconiosis Act No. 57 of 1956, extended the compensation system to asbestos and coal mines and, in that sense, it was progressive. Its principal effect was to remove pulmonary disability from the statute books. The Silicosis Medical Bureau was restructured, and its post-mortem data continued to be hidden from view. Not surprisingly, the new Act saw the compensation rates fall.

[29] Schepers, transcript of evidence before the Beyers Commission: 549.
[30] Schepers, transcript of evidence before the Oosthuizen Committee: 918.
[31] Schepers, transcript of evidence before the Oosthuizen Committee: 934, 936.
[32] Schepers, transcript of evidence before the Oosthuizen Committee: 969.
[33] Schepers, transcript of evidence before the Oosthuizen Committee: 970.
[34] Schepers, transcript of evidence before the Oosthuizen Committee: 972.

Many years later Schepers reflected on that outcome: 'Thereafter pernicious influence aggravated the matter, as the government implemented its fascist doctrines (apartheid) under the guise of racial segregation.'[35] During his term as Director of Saranac Laboratories, Schepers gave a paper at a Toronto conference on occupational disease and medical ethics in which he addressed the dilemma he had faced in Johannesburg in the 1950s: 'How can a physician be effective in bringing about work place reform? What should he do if his good advice is constantly rejected?' [36] He noted that occupational physicians had to reconcile their responsibilities to their patients with their obligations to their employers. By remaining neutral a doctor might preserve his integrity but render himself ineffectual. Schepers reminded his audience that all too often improvements in the workplace only follow litigation or trade union intervention. Under apartheid neither of those avenues was open to the majority of gold miners.

The significance of the Oosthuizen enquiry and the fate of Gerrit Schepers are found in the Pulmonary Disability Act of 1952 which led to a sharp rise in the number of initial certifications of white and black miners. For whites in 1953 there were 306 initial awards, and by 1955 that had grown to 1,466. In 1953 there were 1,748 initial awards to black miners; in 1954 that had risen to 3,099, and in 1956 there were 4,922.[37] Between 1951–52 and 1955–56 the costs of benefits paid to white miners more than doubled, and those to black miners rose threefold. For white miners benefits in 1951–52 cost £1.197 million; in 1955–56 the cost was £2.7 million. The benefits awarded to blacks in 1951–52 totalled £330,000, and in 1955–56 they had risen to £991,000. Those increases were immediately reflected in the levy, which rose from £1.7 million for 1951–52 to £2.4 million for 1955–56.[38] With the passage of the Pneumoconiosis Act in 1956 the number of awards returned to their previous levels as if by magic. Those shifts suggest that the official rates of silicosis and tuberculosis were more a political artefact than a measure of the amount of compensatible disease.

The other face of Sarel Oosthuizen

The Oosthuizen report gave the industry what it wanted with regard to the Pulmonary Disability Act and Gerrit Schepers. For reasons that are not at all clear it also presented a damning critique of the industry's manage-

---

[35] Schepers, transcript of evidence before the Oosthuizen Committee: 971.

[36] Letter from G. W. H. Schepers to Geoffrey Tweedale, 10 July 2000, in author's possession.

[37] G. W. H. Schepers, 'Residual Problems Relating to the Pneumoconioses' paper presented to the Eighth Conference of the McIntyre Research Foundation on Silicosis, Toronto, Ontario, Canada, 22–24 October 1956.

[38] *Report of the Pneumoconiosis Board for the Year Ended 31st March, 1957* (Pretoria: Government Printer, 1957): 11, 15.

ment of tuberculosis and in particular its repatriation policies. That critique makes the report one of the most important reviews of miners' health produced during minority rule. It was certainly the most ambiguous.

The Oosthuizen inquiry was the first since the Medical Commission of 1912 to use the synergy between tuberculosis and silicosis as its starting point: 'There seems to be no doubt that tuberculosis is the most common complication of silicosis in Native labourers and that the problem is considerably more serious than is generally realised.'[39] The Committee found that the official data grossly underestimated the actual disease rates. Part of the problem, it said, was the Silicosis Medical Bureau's total reliance upon sputum testing for diagnosis. A more adequate approach would 'have revealed a more marked connection of the two diseases [silicosis and tuberculosis] in Native labourers.' In addition, the official returns on the total complement of black labour on scheduled mines, included surface workers thereby making them virtually worthless as a guide to the disease rates.[40] The post-mortem returns from the SAIMR were unreliable because so very few black miners certified with silicosis in the second or third stages were examined.[41] Presumably many died from tuberculosis, but their deaths went unrecorded. 'The complete lack of reliable statistical data' Oosthuizen wrote, 'is a matter for grave concern because without the aid of proper statistics, research becomes a wasted effort.'[42]

The Committee noted that black miners were more heavily exposed to silica dust than were whites and therefore were more likely to be affected.[43] Job reservation which had always been a feature of the mines guaranteed whites promotion away from dusty work while preventing the upward mobility of blacks. Prior to 1986, the position of team leader (or 'boss boy', as it was known) was the best job available for a black miner. Team leader work was at the stope face, where the dust exposures were always high.[44] Black miners could never escape the dust.

Oosthuizen quoted at length from a 1951 study of blacks who had ceased working and were subsequently examined by the Bureau. Out of 1,101 surviving miners, 574 had silicosis and a further 71 had silicosis and tuberculosis. Out of 72 deceased subjects, 5 had silicosis and 37 silicosis with tuberculosis.[45] Those results suggested that silicosis was being

---

[39] *Report of the Silicosis Board for the Period 1st April, 1952 to 31st March, 1956* (Pretoria: Government Printer, 1956): 7.

[40] *Report of the Departmental Committee of Enquiry into the Relationship between Silicosis and Pulmonary Disability and the Relationship between Pneumoconiosis and Tuberculosis, Part 2, The Relationship between Pneumoconiosis and Tuberculosis,* Departmental Committee to Inquire into the Definition of Silicosis and Chest Diseases, Departmental Committee 1954, South African National Archives F 33\671 Treasury: 29.

[41] *Report of the Departmental Committee of Enquiry, Part 2*: 26–27.

[42] *Report of the Departmental Committee of Enquiry, Part 2*: 24.

[43] *Report of the Departmental Committee of Enquiry, Part 2*: 30.

[44] *Report of the Departmental Committee of Enquiry, Part 2*: 71.

[45] See Trapido, 'The Burden of Occupational Lung Disease': 23–24.

underdiagnosed and that the longer a miner had silicosis the more likely he was to develop tuberculosis. Oosthuizen also cited data submitted by Dr Becker of the Silicosis Bureau. The data showed that during the period 1939 to 1952 four out of every five white miners who contracted tuberculosis also had silicosis[46] and that at autopsy 80 per cent of black miners showed evidence of tuberculosis.[47] The two diseases were inseparable.

The Oosthuizen Committee was especially critical of the quality of the entry examinations, noting that cases of tuberculosis had been diagnosed in men after as little as a month's service. 'The type of clinical initial examination which Native recruits for scheduled mines receive at present is not worth the time and energy, which is very little, spent on it and might as well not be carried out.'[48] While employers objected to periodic X-rays on the grounds of cost, the Committee argued that they were essential to control disease.

The final medical examinations prescribed under the Silicosis Act were judged of little value: 'There is reason to believe that a fair number of Natives develop active tuberculosis soon after leaving the mines.'[49] That was borne out by the high percentage of certifications in former miners who lodged claims for compensation. In 1950–51, out of 1,173 former miners examined by the Bureau, more than 71 per cent had a compensatible disease. 'It is unlikely that the majority of these Natives developed silicosis after having left the mines and the implications are that they were not certified by the Bureau when they were discharged from the mines. It tends to illustrate the unsatisfactory manner in which Native labourers are examined on discharge.'[50] Oosthuizen recommended that an X-ray be part of the final examination.[51] It also recommended that to reduce the risk of infection in the compounds the number accommodated in each room be decreased from twenty to fourteen,[52] that miners receive an X-ray after six months' employment and thereafter compulsory annual radiological and clinical examinations irrespective of their length of service.[53] To encourage miners to submit to treatment at the end of their contracts they should be paid full wages for a reasonable period while they were being treated.[54]

The most disturbing parts of the Oosthuizen report are those dealing with the 'death trains'. It estimated that approximately seven hundred black miners with active tuberculosis were repatriated to the rural areas annually. The scale of repatriations to adjacent colonial territories was

[46] *Report of the Departmental Committee of Enquiry, Part 2*: 28.
[47] *Report of the Departmental Committee of Enquiry, Part 2*: 30.
[48] 'Memorandum submitted by Dr B. J. P. Becker, Silicosis Medical Bureau' in *Report of the Departmental Committee of Enquiry, Part 2*: 8.
[49] *Report of the Departmental Committee of Enquiry, Part 2*: 122.
[50] *Report of the Departmental Committee of Enquiry, Part 2*: 126.
[51] *Report of the Departmental Committee of Enquiry, Part 2*: 29.
[52] *Report of the Departmental Committee of Enquiry, Part 2*: 126.
[53] *Report of the Departmental Committee of Enquiry, Part 2*: 129.
[54] *Report of the Departmental Committee of Enquiry, Part 2*: 109.

unknown.[55] At that time over 50 per cent of mine labour was recruited from outside South Africa:

> The position at present is most unsatisfactory. Sick Natives are all repatriated through the W.N.L.A. Hospital and it appears that the criterion for deciding whether they are fit for repatriation is fitness to travel, the measurement of which is the ability to stand. It then also seems that it was not an uncommon occurrence for Natives to die in the train on their way home and it appears that whatever control there is at present, unsatisfactory as it may still be, was introduced after vigorous objections had been lodged by the Railway authorities.[56]

It was inhumane, Oosthuizen remarked, to send infected men back to the reserves and allow them to spread tuberculosis to their families and to die.[57] Tuberculosis was rapidly assuming epidemic proportions in the Union and unless steps were taken to put a stop to the repatriation of infected miners, a tragic state of affairs would soon arise.[58] Oosthuizen recommended that it should be an offence for a mine to discharge an employee suffering from active tuberculosis.[59]

The outcome

Following the completion of the Oosthuizen enquiry, the Secretary for Health wrote to the Chamber. He noted that since recruits were given a pre-employment medical examination in the majority of cases they must be contracting tuberculosis on the mines. If those men were treated successfully, it would benefit the communities from which the industry recruited its labour. The Chamber's legal adviser, B. T. Tindall, was unsympathetic. Many migrant workers in other industries returned to the Territories with infective tuberculosis. Therefore treatment by the mines alone would not halt the spread of disease. There was also need for a complementary state scheme in the Territories. Tindall argued that since gold mining was the only industry responsible for paying compensation for tuberculosis, it would be unfair of government to single out the mines for further expenditure. In his view, the industry should not treat black miners.[60]

In October 1954 the Group Medical Officers met to discuss Oosthuizen. The meeting was chaired by Orenstein and included Retief, the WNLA's senior medical officer and Tindall. In contradiction of a century-old

[55] *Report of the Departmental Committee of Enquiry, Part 2*: 132.

[56] *Report of the Departmental Committee of Enquiry, Part 2*: 131.

[57] *Report of the Departmental Committee of Enquiry, Part 2*: 133.

[58] *Report of the Departmental Committee of Enquiry, Part 2*: 131.

[59] *Report of the Departmental Committee of Enquiry, Part 2*: 142.

[60] Memorandum from B. T. Tindall, Legal Adviser to the Manager, Transvaal and Orange Free State Chamber of Mines, Johannesburg, 17th July, 1954, WNLA 20L Diseases and Epidemics, Tuberculosis March 1953–Nov. 1954, Teba Archives, University of Johannesburg: 1.

medical orthodoxy, the Group asserted that even in severe cases there was no conclusive evidence that pulmonary tuberculosis in miners was necessarily associated with silicosis. They recommended that the Chamber go no farther than to acknowledge that the presence of silica in the lungs favoured the development of infection. The Group argued that the exit examinations were adequate, citing as evidence the fact that over the previous year only three former miners in the Native territories had been compensated.[61] Orenstein suggested that as a compromise the industry might agree to an X-ray at the final examination for black miners with two years' continuous service. In practice the scheme would be limited to East Coasters and Tropicals. Two weeks later the Group Medical Officers met again to prepare a statement of evidence. Under Orenstein's guidance it concluded that the incidence of tuberculosis uncomplicated by pneumoconiosis was lower among blacks employed on the mines than among the general population – in other words, the flow of disease was from the rural areas to the mines.[62] Oosthuizen was wrong.

By the time the Group Medical Officers had decided there was no tuberculosis problem Gerrit Schepers had already begun a new life in the USA. He never again worked in South Africa but he did over the coming decades make a number of visits home. In January 1963 Schepers gave a paper to a general meeting of the MWU in Johannesburg. He told his audience that almost 50 per cent of black miners would eventually die of tuberculosis while a large number would also contract silicosis. Because the mines could not operate without migrant labour, he said, the MWU had a moral responsibility to ensure that black miners, who had no means to protect themselves, were protected.[63] The MWU chose not to take up that challenge.

---

[61] Minutes of a Meeting of the Sub-Committee of Group Medical Officers held on Thursday, 7th October, 1954, in the Chamber of Mines Building, WNLA Diseases and Epidemics, Tuberculosis March 1953–Nov. 1954, Teba Archives, University of Johannesburg.
[62] Minutes of the Meeting of the Sub-Committee of Group Medical Officers held on Monday, 25th October, 1954, in the Chamber of Mines Building, WNLA 20L Diseases and Epidemics, Tuberculosis March 1953–Nov. 1954, Teba Archives, University of Johannesburg.
[63] Dr G. W. H. Schepers, transcript of speech to the Mine Workers' Union, Johannesburg, Friday, 25 January 1963, in author's possession.

8

Healing Miners

The period from 1954 to 1980 saw the industry faced with two major challenges: a change in the pattern of recruitment and the introduction of treatment for pulmonary tuberculosis. Migrant labour was the life blood of the mines, and the Chamber of Mines had at various points in its history responded successfully to labour shortages. A cure for tuberculosis was something the mines had never faced. Apartheid would eventually prove the perfect setting for resolving that challenge to the industry's advantage.

After World War II the gold mines began to lose South African workers to better-paid jobs in other industries. By 1960, Malawi and Mozambique were supplying the bulk of mine labour. Ten years later only 25 per cent of black gold miners were South African, the rest being from Lesotho (29 per cent), Southern Mozambique (21 per cent) and the tropical North (24 per cent) – mainly from Malawi.[1] Three factors led to a reversal in that pattern. The freeing of the gold price in 1973 initiated a period of expansion and enhanced profits for the industry as a whole. In May 1974 President Banda of Malawi withdrew all Malawian labour after an aeroplane crash killed miners in transit to Johannesburg. Finally, the revolution in Mozambique saw a dramatic drop in recruitment from that country. Deprived of labour from those sources, the mines used wage increases to attract workers from the Eastern Cape. Increases in average mine wages by 62 per cent in 1974 and a further 68 per cent in 1975 saw the number of local recruits rise sharply. Early-return bonuses to shorten the time spent away from the mines were also introduced between 1977 and 1985.

The discovery of streptomycin and then isoniazid saw the first effective treatment for tuberculosis become available in the early 1950s. Patients were given streptomycin daily for three months and isoniazid and thiacetazone for eighteen months.[2] Treatment took two years to complete during which time the patient had to remain under medical supervision. (After

---

[1] Trapido, 'The Burden of Occupational Lung Disease': 28.
[2] R. L. Cowie, 'Pulmonary Tuberculosis in South African Gold Miners: Determinants of Relapse after Treatment' Master's thesis, McGill University, 1987: 5.

the introduction of rifampicin in 1967, short-course regimens of between six and twelve months were developed.) The mines had been repatriating tuberculosis cases since before World War I, and both the Stratford Commission in 1943 and the Oosthuizen Committee in 1954 found that the system was a threat to public health. The availability of effective treatment did not solve the problem of what to do with sick miners.

Repatriation was cheap, and it meant that most miners with tuberculosis died in rural areas. Prolonged treatment was expensive, and the Chamber feared that by making disease more visible it would have a negative impact on recruitment. The Gold Producers Committee (GPC) devoted much of its time to that issue yet, remarkably, the treatment of tuberculosis in migrant workers was not discussed at the international conference held at Johannesburg in 1959. Neither did it feature at the South African commissions of inquiry into occupational health of 1964 and 1976.

The industry wanted to minimise the cost of managing tuberculosis. Treatment had to be cheap, and the best treatment of all would allow miners to continue working while undergoing chemotherapy. The possibility that men who had been successfully treated could return to work had been raised by the Mine Workers Union (MWU) at the Oosthuizen inquiry. White miners, once they left the industry, had difficulty in finding comparable employment, and the MWU wanted its members to be allowed to return underground. All the experts consulted by Oosthuizen, including UK and US authorities, strongly opposed reemployment.[3] When the Group Medical Officers met in October 1954 to discuss the Oosthuizen report, the Chamber's legal adviser raised the issue of reemployment. Experiments carried out in the US on workers with healed tuberculosis showed that in most cases a man who returned to dust work relapsed.[4] Those studies convinced the subcommittee that reemployment – or, as it was later to be termed, reactivation – was inadvisable. The work of Dr R. L. Cowie at the Ernest Oppenheimer Hospital at Welkom would eventually see that decision overturned.

Negotiations over treatment

The new-generation drugs offered a cure and the industry was soon under pressure from the Department of Health to revise its policies. Because treatment took so long there was the problem of where the men would be housed and who should pay. In August 1954 the Group Medical Officers' subcommittee considered a request from the Secretary for Health that the mines provide treatment.[5] The Secretary pointed out that a number of major employers in the Cape and Natal had agreed to treat black workers

[3] *Report of the Departmental Committee of Enquiry, Part 2*: 115.
[4] Minutes of a Meeting of the Sub-Committee of Group Medical Officers, 7th October, 1954: 2.
[5] Minutes of Meeting of the Sub-Committee of Group Medical Officers held in the Chamber of Mines Building at 2:30 p.m. on Tuesday, 3rd August, 1954, WNLA 20L Diseases and Epidemics, Tuberculosis March 1953–Nov. 1954, Teba Archives, University of Johannesburg.

until they were no longer infective and he suggested that the mines introduce a similar scheme. In principle the Group Medical Officers supported
the proposal, but there were a number of problems to be resolved. They
included the lack of accommodation at the Witwatersrand Native Labour
Association (WNLA) hospital and the management of patients detained at
individual mines. Both would require major investment and the medical
officers insisted that the cost be borne by the state.[6]

In September 1954 senior officers from the Department of Health
attended a conference at the Chamber's offices. During a lively discussion
Dr van Rensburg of the Department pointed out that miners were often
only diagnosed when they were seriously ill and infective, supporting his
claim with data showing that 27 per cent of repatriated cases had positive
sputa. James Gemmill explained that all such cases were given streptomycin and isoniazid at the WNLA hospital for an average of six weeks
while awaiting compensation.[7] Once they were repatriated, the WNLA
had no further legal obligation, and it was assumed that treatment would
be continued in the patient's home district.

One of the factors contributing to the spread of tuberculosis was the
definition of the disease in the Silicosis Act No. 47 of 1946. Under the Act
a mine worker was deemed to be suffering from tuberculosis if there was
evidence of tubercle bacilli in his sputum or if he had serious impairment.
Many miners with tuberculosis had neither. The Beyers Commission had
been critical of the law, noting the importance of removing infected men
to protect their fellow workers. Beyers found that in many cases the definition under the Act prevented the Bureau from correctly certifying sick
men. Oosthuizen agreed: in active and advanced cases there might be no
positive sputum and no marked disability. It recommended that a person
be deemed to have tuberculosis if *manifestations of the disease were
present* irrespective of whether the capacity for work had been impaired.[8]
That recommendation was incorporated into the Pneumoconiosis Act No.
57 of 1956, which also required every black miner to be X-rayed prior to
employment, medically examined at regular intervals while employed,
and examined again at discharge.

In June 1956 the Department of Health asked the Chamber to introduce
full-sized X-rays for black miners. The Group Medical Officers considered
then rejected the proposal on the grounds of cost. In any case, they saw
little advantage in such an initiative.[9] In their view, the entry examination,

[6] Circular No. 277/54, from A. J. Orenstein, Chairman, Sub-Committee of Group Medical Officers, Gold Producers' Committee (Internal), 19th August, 1954, WNLA 20L Diseases and
Epidemics, Tuberculosis March 1953–Nov. 1954, Teba Archives, University of Johannesburg.
[7] Conference held in Office of Mr J. A. Gemmill on the 15th September, 1954, WNLA 20L
Diseases and Epidemics, Tuberculosis March 1953–Nov. 1954, Teba Archives, University of
Johannesburg.
[8] *Report of the Departmental Committee of Enquiry, Part 2*: 78–82.
[9] Minutes of Meeting held 25 June 1956 of the Sub-committee of Group Medical Officers, Gold
Producers' Committee, Johannesburg, NRC P2A Miners' Phthisis Compensation to Natives,
Teba Archives, University of Johannesburg.

which involved a miniature X-ray, was sufficient. At its Welkom mines in the Free State, Anglo American had used mass miniature X-rays at entry with good results. The official tuberculosis rate was 0.127 per cent as against 0.444 per cent on the Witwatersrand. In the Free State the silicosis rate was 0.024 per cent, while the Rand figure was far higher at 0.197 per cent.[10] That was not, however, the end of the matter. The Department of Health was determined that the mines introduce treatment, and between August 1956 and March 1958 it convened a series of conferences with industry to promote such a scheme.

Dr Clarke, from the Department of Health, chaired a conference on tuberculosis in August 1956, during which the Chamber made it clear its position on treatment. The mines, it argued, were already notifying the Department of repatriated cases so that they could be followed up and 'any other step, such as a proposal to send such cases to sanatoria, either on the Reef or elsewhere, would have an adverse effect on recruitment'.[11] During what proved to be difficult negotiations there were also meetings between the Chamber, the Department of Mines and the Silicosis Medical Bureau about the conduct of medicals. The major point of dispute was the periodic examinations. The Department of Mines wanted six-monthly exams involving both an X-ray and clinical review.[12] As always, the issue was cost. Periodic examinations had to be carried out when miners were coming off shift. On mines with five thousand or more employees the processing of at least two hundred men at a time would take over two hours and a clinical exam even longer.[13] The Chamber estimated the primary cost including lost shifts, additional staff and plant to be £400,000 per year. Disruption to work schedules would cost a further £800,000 annually in lost production.[14] According to the Chamber such an initiative was unnecessary. The high rates of injury and disease meant that around 50 per cent of black miners were admitted each year to hospital and when it was deemed necessary they were X-rayed. The Chamber made no mention, however, of a second problem: the use of X-rays may have then identified previously undiagnosed disease.

In September 1957 the Minister for Mines, A. J. van Rijn, convened a round table conference with the Chamber and the MWU to discuss the

---

[10] Memorandum: Gold Producers' Committee, the Pneumoconiosis Act and Medical Examination of Native Labourers, 21 August 1956, NRC 637/7 Miners' Phthisis Compensation to Natives 1955–1958, Teba Archives, University of Johannesburg.

[11] Notes of a Conference on Tuberculosis: Native Mine Labourers and the Act 57 of 1956, Pretoria, 29 August 1956, NRC 637 7 Miners' Phthisis Compensation to Natives 1955–1958, Teba Archives, University of Johannesburg.

[12] Memorandum: Gold Producers' Committee, the Pneumoconiosis Act and Medical Examination of Native Labourers.

[13] Memorandum: Pneumoconiosis Act: Medical Examination of Native Labourers by the Transvaal Chamber of Mines, 9 August 1956, NRC 637 7 Miners' Phthisis Compensation to Natives 1955–1958, Teba Archives, University of Johannesburg.

[14] Memorandum: Gold Producers' Committee, the Pneumoconiosis Act and Medical Examination of Native Labourers.

industry's future. The Chamber was concerned about the impact of the levy on profitability. Many mines were reaching the end of their lives, and some were facing imminent closure. Daan Ellis of the MWU noted that the compensation levy had risen from £1,800,000 in 1955 to over £4,000,000 in 1957. He was concerned that the expense of repatriating black miners would cost white jobs. He wanted 'proper pre-employment medicals' to keep infected blacks out of the mines. Apart from his concern about the 'enormous sums' paid in compensation, he was convinced that the health of white miners and their families was at risk.[15] The director of the Pneumoconiosis Bureau, Dr J. Loots said that the number of claims by whites was falling but there had been a rise for black miners due to changes in the definition of tuberculosis. He estimated that of the black miners certified two-thirds were recruits, the majority of whom had previously worked underground.[16]

The negotiations between the Chamber and the Departments of Health and Mines dragged on. The industry had a number of concerns. If recruits were to be detained until they were no longer infective, extra accommodation would be required. Extended periods of treatment during which miners received no pay would have an adverse effect on recruitment. After much discussion the industry agreed that treatment should be limited to six weeks. On their return home patients would undergo domiciliary care by state medical officers for a period of up to six months.[17] A patient's X-rays, together with a copy of his hospital records, was to be forwarded by the chief regional health officer to the Bantu Affairs office in the miner's home district. Under the Public Health Act, local authorities were to ensure 'that adequate measures are taken for preventing the spread of the disease, including provision for his accommodation, maintenance, nursing and medical treatment'.[18] I have found no evidence that such a referral procedure ever existed.

A scheme for the treatment of South African miners at the WNLA depots who were awaiting settlement of their compensation claims was eventually introduced in July 1958. Medication was provided for a maximum of six weeks, and the costs were borne by the Department of Health at a rate of R1.40 per man per day. From March 1960 the scheme was extended to non-Union labour. Treatment was carried out at mine hospitals, and once again the costs were borne by the Department. In principle the Department reserved the right to decide when a patient should be discharged, but there is no evidence that such an authority was exer-

---

[15] Report of Round Table Conference Convened by the Minister for Mines, A. J. van Rijn, to Discuss Vulnerable Mines, 13 September 1957, NRC 637 Miners' Phthisis Compensation to Natives 1955–1958, Teba Archives, University of Johannesburg: 3, 9.

[16] Report of Round Table Conference: 4–5.

[17] Mr B. T. Tindall, Legal Adviser, 'A Brief History', appendum to letter from A. T. Milne, General Manager, to Members of the Gold Producers' Committee, 4th April, 1962, Johannesburg, re: Native Labour: Payment during Treatment in Mine Hospitals, WNLA 10/1 Hospital Charges, Health Services August 1958 to September 1962: 5.

[18] Tindall, 'A Brief History': 6.

cised.[19] As always, control remained in the hands of mine medical officers.

There was no provision for the ongoing treatment of non-Union miners. Such patients normally received chemotherapy at the WNLA while their cases were being assessed or until they became non-infective. According to B. T. Tindall of the Chamber:

> Apart from the notification made to the Authorities in Nyasaland no notification is made to the Authorities of other tropical areas or to the Authorities in Swaziland on the repatriation to these areas of Natives found to have contracted tuberculosis. I am extremely reluctant to recommend to the Gold Producers' Committee that the Industry should take upon itself an obligation which at this stage falls clearly upon the local authorities ... if this will involve the Industry in considerable additional expenditure.[20]

The impact of that policy was felt by South Africa's neighbours. In 1957 the annual Public Health Report for the newly created Federation of Southern and Northern Rhodesia and Nyasaland noted that despite radiography programmes tuberculosis was becoming a more serious problem.[21] In 1960 more than half a million pounds was spent on care and prevention. Two years later eight hundred beds in government and mission hospitals were occupied by tuberculosis patients. Around 30 per cent of those treated as in-patients absconded and of that group fewer than 30 per cent returned for treatment. Federation health authorities feared that drug-resistant strains of infection would soon develop.[22]

The Chamber was in a bind. It had never accepted tuberculosis as an occupational disease, insisting instead that infection was brought to the mines from rural areas. According to The Chamber the failure of government to treat a rural epidemic was spreading infection and depriving the mines of labour. Even if the state paid for the cost of medication, the industry was left with providing beds for men who, having been treated, were forbidden by law to work in a mine. The obvious solution was the reemployment of miners who had been treated.

As early as 1961 the Native Recruiting Corporation (NRC) was discussing reactivation. In October of that year the NRC's General Manager complained that the inclusion of tuberculosis as a compensatible disease was having a negative impact on public opinion. 'In principle there can be no more objection to Natives who have been cured of pulmonary tuberculosis being given employment on the mines than there can be to their being given employment anywhere else. It is well known that pulmonary

---

[19] Tindall, 'A Brief History': 4.

[20] Tindall, 'A Brief History': 8.

[21] *Annual Report on the Public Health of the Federation of Rhodesia & Nyasaland for the Year 1957* (Salisbury: Government Printer, 1958): 9.

[22] *Annual Report on the Public Health of the Federation of Rhodesia & Nyasaland for the Year 1959* (Salisbury, Government Printer, 1960): 10, 12, 14.

tuberculosis is not an occupational disease peculiar to mining.'[23]

In April 1962 the Chamber's General Manager, A. T. Milne, canvassed the views of the mining houses, senior medical officers and the Department of Health. There was general agreement that the existing arrangements were unsatisfactory, as most patients at mine hospitals failed to complete treatment. The Rand Mines chief medical officer, Dr A. M. Coetzee, commented: 'The men complain that their families suffer if they stay in hospital without receiving any income. If we use any coercion, a riot threatens. If the men are not treated satisfactorily before repatriation, the whole scheme fails.' Patients became anxious about their families and left hospital as soon as they felt better. Consequently, many returned home while still infective. The provision of sick pay for miners under care would, Milne said, improve the situation.[24] The Secretary of the Mine Medical Officers' Association (MMOA) agreed: 'It is common experience in mine hospitals that long-term native patients ... become restless and discontented and agitate to be released from hospital ... This is most undesirable ... and my Executive Committee therefore suggests ... some form of subsistence allowance being provided.'[25] James Gemmill suggested that while in treatment such men should be employed in surface occupations and receive pay. 'The work they performed would, of course, have to be in non-dusty occupations. I also think it inadvisable for the Industry to attempt to compel these patients to remain for the full period, as this would inevitably have an adverse effect on the popularity of mining.'[26]

The Gold Fields Company was unwilling to pay for treatment but agreed to contribute to an *ex gratia* scheme. Coetzee agreed that it was difficult to keep men in hospital and suggested that patients be given surface work. Gemmill noted that if tuberculosis patients were paid while convalescing there would be discontent from men recovering from other ailments who received no such benefits. Milne was concerned that forcing men to receive treatment over some months might have an impact on recruitment. He rejected the suggestion that the mines should enforce treatment. That would involve additional expense, and besides, the responsibility lay with government not the mines.[27]

By April 1962 the system of treatment had been in operation for almost four years. In principle black miners were treated until they ceased to be infectious and then sent home to continue their treatment at local hospitals. But such facilities were limited and where hospitals or clinics did exist former miners were reluctant or unable to complete chemotherapy. Recruits from outside South Africa were treated until non-infectious and

---

[23] Letter from A. T. Milne, General Manager, to Members of the Gold Producers' Committee, 4th April, 1962, Johannesburg, re: Native Labour: Payment during Treatment in Mine Hospitals, WNLA 10/1 Hospital Charges, Health Services August 1958 to September 1962: 2.

[24] Letter from A. T. Milne: 2.

[25] Letter from A. T. Milne: 2–3.

[26] Letter from A. T. Milne: 3.

[27] Letter from A. T. Milne: 8.

then repatriated. In case of Nyasa the local public health authorities were notified. No notification was given to health departments in Basutoland or Mozambique.[28]

The WNLA hospital and repatriation

The conditions at the WNLA hospital in Johannesburg, which lay at the centre of the repatriation system, provide the context for the negotiations between the Chamber and the Department of Health. Fortunately, we have a first-hand account of what the hospital was like from a doctor who spent his career with the WNLA. Dr Oluf Martiny was born in the Orange Free State in 1924 and educated at Smithfield and at the University of the Witwatersrand, where he graduated in medicine in 1949.[29] He worked briefly in Denmark and then for two years at the Missouri State Sanatorium in the USA, where he treated tuberculosis and silicosis patients, many of them stonemasons. Martiny returned to South Africa and joined the WNLA in 1954. He remained with the company until his retirement thirty years later.

The WNLA hospital and the adjoining compound were a transit centre for recruits on their way to and from the mines. The hospital also processed miners who were being repatriated. There was a massive flow of men through the system. The hospital had twelve hundred beds, including a new wing that was completed shortly after Martiny started work. The hospital was overcrowded and unhygienic, and the accommodation for the black staff living on site was inadequate. James Gemmill and the GPC controlled the budget and decided what would be done medically.[30] Gemmill was determined to keep costs down, and the system was run as cheaply as possible. The staff were poorly paid, and the conditions for patients were deplorable.

Martiny was put in charge of the tuberculosis and chest diseases wards. Every morning, the WNLA doctors would start their day by examining recruits, who were lined up naked. The examinations were the most important part of the day's work as far as management was concerned. On the busiest morning Martiny and his six colleagues examined twelve thousand men.[31] Martiny also did reviews of medical stations in the rural areas. At the beginning of each week there was an influx of recruits and experienced miners who had completed their contracts. They were housed in cramped rooms, with four or more sleeping in double bunks, and a coal stove in the middle for cooking and warmth. The men slept side by side with strangers on cement platforms. Most preferred to sleep outdoors, even in winter, on the dirty tarmac, where there were no bedbugs and the

---

[28] Letter from A. T. Milne: 7, 8.
[29] Interview with Dr Oluf Martiny, Forest Town, Johannesburg, 27 April 2011.
[30] Interview with Dr Oluf Martiny.
[31] Oluf Martiny, 'My Medical Career' MS, Johannesburg, September 1999: 7.

air was clean. Pneumonia and influenza were common.[32]

The hospital wards were filled with tubercular miners and recruits who had been diagnosed on their arrival at Johannesburg. There were also men with extreme silicosis, some of whom had worked underground for as little as six months.[33] The men would stay while their service record was verified for the purpose of compensation, which usually took six weeks. While waiting for the Bureau's decision, the WNLA doctors began treatment. On discharge each patient was given a letter of referral to his nearest clinic.[34] Most were fit to travel, but once they left the WNLA there was no treatment. In Martiny's words, 'they were doomed'.[35]

From its inception the repatriation system was one of the most contentious aspects of mine medicine. In addition to miners with lung disease, there was a steady flow of men who had been seriously injured in accidents. The treatment of those men is a measure of the medical system. Trains carrying injured or sick miners left the Booysens Railway station twice each week. In June 1951 the district manager for the WNLA in Mozambique wrote to head office about two paralysed miners who had been repatriated to Ressano Garcia. Both men had worked six or seven contracts, and their weeping families were upset that they had received no compensation.[36] Two Portuguese farmers who were present at the railhead when they arrived made disparaging comments about the WNLA. The District Manager suggested that some form of *ex gratia* payment would help to prevent further criticism.

In April 1952 the same officer wrote again to the head office about the need to improve the management of disabled miners. Seriously injured men arriving at Lourenço Marques by train were carried on stretchers through the town in the back of a camping van and then dispatched by ship to their home villages. That spectacle had aroused unfavourable comment from local authorities. The Manager suggested that the WNLA purchase a second-hand ambulance or at least a panel van fitted with stretchers mounted on springs to transport the men to the boat.[37] Those cases are only unusual because they generated a correspondence. Most injured miners, many of whom were permanently disabled, were moved almost invisibly though the system. On 15 October 1959 the WNLA repatriated forty-seven men. Listed at random, their injuries included a fractured skull, an amputated right femur, a fractured left foot, an amputated left leg, a fractured left tibia, a fractured pelvis, a fractured right tibia and fibula, a fractured left foot, a fractured spine and a fractured tibia and fibula. There were also cases of tuberculosis of the spine, lobar pneu-

[32] Martiny, 'My Medical Career': 19.

[33] Interview with Dr Oluf Martiny.

[34] Martiny, 'My Medical Career': 39.

[35] Interview with Dr Oluf Martiny.

[36] Letter from the District Manager, WNLA, Lourenço Marques, 28 June 1951, to the General Manager, WNLA, Johannesburg, 22 April 1952, WNLA 35 Repatriations and Rejects: East Coast Natives March 1925 to August 1957, Teba Archives, University of Johannesburg.

[37] Letter from the District Manager.

monia, enlarged liver and spleen, blind left eye and deafness.[38] In a tight labour market even a small physical incapacity could result in a total loss of employment for an unskilled worker.[39]

The conditions on the WNLA wards were wretched. There were patients in the beds and wherever they could find a space to sleep. Many lay on the cement floors with a thin felt pad for a mattress and a single blanket. During the daytime they would sit under the trees in their night shirts. At night the wards were filled with the sound of productive coughing and spitting.[40] Although hundreds of men were processed each week, the system was cheap to operate. During 1956 the cost of repatriating compensated black miners was £13,000, and a further £6,500 was spent on repatriating men who were unfit to work but had not received compensation.[41]

Soon after joining the WNLA, Martiny began a study, tracking tuberculosis cases sent home over the previous ten years. The medical facilities in the rural areas were virtually non-existent. At Libode in the Transkei the local doctor provided treatment for only a month. Patients who had improved but needed ongoing care were discharged to make way for new admissions. Many of those patients relapsed and were subsequently readmitted. Rural hospitals and clinics had to wait weeks or months before they received fresh supplies of medication. Those delays interrupted treatment and led to drug resistance. Martiny told the Oosthuizen Committee in 1954 that nearly 90 per cent of men repatriated with tuberculosis had discontinued their treatment.[42] In Martiny's opinion, the entry and exit medical examinations were inadequate for identifying all but advanced disease, the tuberculosis rate was far higher than the Bureau data indicated and mass miniature X-rays were missing many cases of infection.

The flaws in the medical review system were discussed at length by the GPC. In 1957 it was not possible to take X-rays at every point of recruitment: many of the labour-sending areas were isolated and the costs would be prohibitive. The GPC estimated the cost of establishing a minimum of twenty clinics at £100,000 and operating them at a further £40,000 a year.[43] As one medical officer later commented during a meeting of the MMOA, 'If stringent examinations were carried out, there would be fewer applicants. In the case of Africans there would be more rejects and consequently more people in the reserves would be unable to make a living.'[44]

---

[38] Letter from Secretaries, Transvaal and Orange Free State Chamber of Mines, to Breyner & Wirth, Limited, Lourenço Marques, 15th October, 1959, WNLA 35 Repatriations and Rejects: East Coast Natives September 1957 to October 1959, Teba Archives, University of Johannesburg.

[39] See R. S. Arkles, A. J. Weston, L. L. Malekla and M. H. Steinberg, *The Social Consequences of Industrial Accidents: Disabled Mine Workers in Lesotho,* National Centre for Occupational Health Report No.13/1990.

[40] Martiny, 'My Medical Career': 19.

[41] Memorandum: Gold Producers' Committee to the Round Table Conference: 7.

[42] Martiny, 'My Medical Career': 40.

[43] Memorandum: Gold Producers' Committee to the Round Table Conference: 5.

[44] See discussion in *Proceedings of the Mine Medical Officers' Association* 43, no. 385 (1963): 36.

Science and risk

The high rate of pulmonary disease at the WNLA hospital did little to stimulate research on migrant workers. In 1954 the Chamber of Mine's operating budget exceeded £1 million for the first time, and its investment in medical research was substantial. Surprisingly little of that money was devoted to lung disease. In 1957 the research budget was £254,000, of which £24,000, or less than 10 per cent, went to silicosis and tuberculosis. In the following year the allocation fell to £18,669.[45] Despite the lack of investment, between 1954 and 1967 a series of studies called into question the effectiveness of medical surveillance. Two of those studies came from the newly formed Pneumoconiosis Research Unit (PRU), and two were commissioned by Anglo American.

With the passage of the Pulmonary Disability Act in 1952, the Silicosis Medical Bureau had purchased pulmonary function equipment for assessing compensation claims. That equipment offered Dr Margaret Becklake and her colleagues at the PRU, L. du Preez and W. Lutz, the opportunity to explore the correlation between the X-rays of silicotic patients and demonstrable disability. It was a problem in the medical literature that Watkins-Pitchford and Watt had identified in 1912. In Becklake's words: 'The lack of correlation between clinical and physiological status on the one hand, and radiological status on the other, has been frequently reported in classical silicosis.'[46] The issue was important for both diagnosis and compensation and the Chamber cooperated by making miners available. Prior to publication the results were reviewed by the Group Medical Officers' subcommittee.[47]

Becklake examined two cohorts: 476 miners with normal chest X-rays and 85 with marked nodular silicosis throughout both lungs. A total of twenty measurements of vital capacity, expiratory flow and maximum breathing were recorded for each case. 'Statistical analysis confirmed the impression gained by scrutiny of the results that no single test [of the twenty used for the study] discriminated satisfactorily between radiologically normal miners and those with advanced radiological silicosis.' The results were clear: 'No good evidence could be found in our study for the use of less than the eight tests mentioned as a screen procedure for silicosis.' Becklake concluded: 'Lung function scores were shown to corre-

---

[45] A. T. Milne, General Manager, GPC, Internal Circular No.2/58 6 January 1958, Chamber's Research Programme 1958, NRC 637 Miners' Phthisis Compensation to Natives 1955–1958, Teba Archives, University of Johannesburg.

[46] Margaret Becklake, L. du Preez and W. Lutz, 'Lung Function in Silicosis of the Witwatersrand Goldminer' Pneumoconiosis Research Unit, Johannesburg, 1956, NRC 637 Miners' Phthisis Compensation to Natives 1955–1958, Teba Archives, Universithy of Johannesburg: 1.

[47] Memo: B. T. Tindall to Members of the Sub-Committee of Group Medical Officers, GPC, 19 September 1957, NRC 637 Miners' Phthisis Compensation to Natives 1955–1958, Teba Archives, University of Johannesburg.

late with length of rock breaking service, irrespective of the radiological findings. This suggests that disability due to mining may be found in men in whom there is no frank roetgenological evidence of dust disease.'[48] At the least the results cast doubt on the efficacy of X-rays as the foundation of medical surveillance. They also suggested that many miners with disability were not receiving compensation. Becklake's study, published in the USA in 1958, had no impact in South Africa.[49]

One of Becklake's colleagues at the PRU, Dr C. B. Chatgidakis, carried out an important survey of cardio-respiratory lesions in deceased miners in 1959. Her material came from the autopsies of 1,020 African gold miners who had died suddenly on the Rand. The causes of death ranged from mine accidents to stabbings, gassings and suicides. Almost 90 per cent of the miners involved were less than forty years old, and all were in apparent good health at the time of death. In those respects the cohort was representative of the black workforce. Each case was examined macroscopically, and in most instances lesions were confirmed histologically. The most important finding was that 91 cases or 9 per cent had acute, active tuberculosis. 'The tuberculosis lesions found in this series were all typical of the re-invasive or re-infective type of tuberculosis.' In addition, silicosis was found in 51 men, 19 of whom also had tuberculosis. The incidence of chronic nonspecific pleurisy was very high, 182 cases or 19 per cent of the group. There were also 29 cases of cardiac lesions.[50] The study showed that despite pre-employment and periodic medicals there was a high rate of active tuberculosis among those serving miners. It also suggested that the mines were discharging a far higher number of infective cases than was being picked up by exit medicals. As with Becklake's research, the paper called into question the centrepiece of the industry's surveillance programme, the use of mass miniature X-rays.

The Chamber was supportive of Chatgidakis' research, and her report was initially reviewed in December 1959 by the GPC. Orenstein then forwarded the document to the Chamber's legal adviser. The Chamber's assistant general manager, H. D. Thomson, recommended the report be further investigated before being submitted to the Group Medical Officers' subcommittee.[51] At that point the documentary trail ends. We know that the study was not published, and there is no evidence that it was discussed outside the Chamber. It is unlikely that Chatgidakis' was the only such case, and the suppression of her work raises the question of how many other studies suffered the same fate.

[48] Becklake, du Preez and Lutz, 'Lung Function in Silicosis': 4, 8.
[49] Margaret Becklake, L. du Preez and W. Lutz, 'Lung Function in Silicosis of the Witwatersrand Goldminer' *American Review of Tuberculosis and Pulmonary Disease* 77 (1958): 400–12.
[50] Dr C. B. Chatgidakis, 'An Autopsy Survey of the Cardio-Respiratory Lesions in 1,020 Cases of Unnatural Deaths in African Gold Miners, PRU' NRC 691 P.2a, Miners' Phthisis Compensation to Natives 1960 to 1962, Teba Archives, University of Johannesburg: 5.
[51] Circular from H. D. Thomson, Assistant General Manager, to Members of the Sub-committee Medical Officers, 2 May 1960, NRC 691 P.2a, Miners' Phthisis Compensation to Natives 1960 to 1962, Teba Archives, University of Johannesburg.

In 1960 Anglo American commissioned one of its medical officers, Dr J. G. D. Laing, to research the tuberculosis problem.[52] His survey, which had the support of the Chamber, was conducted at a Free State gold mine, a gold mine on the West Rand and a colliery. Laing's brief was to determine the number of non-immunes (subjects with little immunity to infection) entering the industry, the presence of carriers in the compounds, the effectiveness of the X-ray system, and the attitude of the Pneumoconiosis Bureau to early certification. While the research was in progress, the Pneumoconiosis Compensation Act of 1962 came into force.

Laing found that the skill of mine medical officers in reading X-rays was the critical factor in the detection of early lesions and therefore in the number of certifications. He also found no evidence that clinical examinations improved the detection of early-stage tuberculosis. The literature indicated that in miners the breakdown rate in so-called inactive lesions might be as high as 30 per cent per year making an accurate diagnosis difficult. His key findings were a higher incidence of tuberculosis in the mining population than was generally supposed, a significant percentage of susceptibles entering the mines, a number of whom were being infected there, and an undetected pool of tuberculosis. Laing attributed that disturbing situation to defects in the X-ray screening system, a lack of protection on the mines against the spread of disease, and the inadequate training of mine medical officers.[53]

Laing divided the deficiencies in X-ray screening into four main elements: equipment, technique, interpretation and organisation. The greatest problem lay with the interpretation of the plates. Most medical officers had little experience of tuberculosis and no experience in interpreting miniature films; they were expected to acquire that knowledge on the job. Most mines X-rayed large batches of employees in a few days to minimise the disruption of shifts and thus reduce the cost. The volume of films taken and the speed with which the interpretation was done precluded the employment of specialist radiologists. In any case employers considered the use of specialists unnecessary. Miniature films required careful reading and where large volumes were examined errors increased. Even experienced interpreters would miss on average almost 40 per cent of the lesions on a miniature film.[54]

In theory, the mining industry provided treatment until a miner was non-infective. The patient was then repatriated and his care devolved to outside agencies. Laing found that the system did not work. The minimum period required to achieve a 'cure' was twelve months and, to be safe, treatment for eighteen months or two years was desirable. It was best to keep patients in hospital until they were non-infective and thereafter

---

[52] J. G. D. Laing, 'An Investigation into Tuberculosis in the Mining Industry,' Transvaal and Orange Free State Chamber of Mines Research Organisation, C.O.M. Reference: Project No.732/64A Research Report No. 80/67 1967: 1.

[53] Laing, 'An Investigation into Tuberculosis': 19, 78, 23, 79.

[54] Laing, 'An Investigation into Tuberculosis': 80, 81, 84, 82.

continue treatment on an out-patient basis. The stakes were high as inadequate treatment could lead to resistant strains.[55] Laing's report was submitted to Anglo American in February 1966. The final discussion on the matter was on 10 April 1968, when a subcommittee supported his recommendation on the need to improve diagnostics and to give mine medical officers additional training in reading X-rays.[56] There is no evidence that any such changes were made.

Dust

In the absence of follow-up studies of miners, the industry and the state fell back upon the official data provided by the Silicosis Bureau. The other pillar used to quantify mine safety was the dust levels. While Laing was conducting his research into medical screening, Anglo American commissioned a review of dust exposures. J. de V. Lambrechts surveyed twenty mines with a total underground working population of seventy-five thousand, or 25 per cent of Rand miners. It was a large-scale undertaking and Lambrechts was confident that his results were representative of the industry as a whole. He was also confident that with one or two exceptions the mines were not disclosing their 'bad' dust counts. According to Lambrechts around thirty-five thousand miners were at any time exposed to an unsatisfactory dust risk. Of that group perhaps four thousand had exposes of more than 1,000 ppcc, a lethal concentration. He concluded: 'The fact that the continued incidence of pneumoconiosis is sometimes associated with an element of "mystery" might have a completely logical explanation in the above observations.'[57]

During his project Lambrechts almost certainly consulted the leading researcher in the field, Derrick Beadle. From 1935 until 1953 Beadle worked for the Chamber where he developed an interest in dust physics. In 1953 he joined Rand Mines Ltd and became chief physicist and head of the industrial hygiene section of the Corner House Laboratories. Beadle was recognised as a world authority and between 1965 and 1970 he published a series of papers on dust and silicosis. In one study conducted in the mid-1950s he sampled a cohort of white miners employed in a variety of occupational categories working on twenty mines. The group had started on the Rand in the period between 1934 and 1938 and had worked at least three thousand shifts.[58] When their medical and work

[55] Laing, 'An Investigation into Tuberculosis': 93.

[56] Cartwright *Doctors of the Mines*: 134.

[57] J. de V. Lambrechts, *Investigation undertaken by Anglo American Corporation of South Africa Limited, An Industry Survey of Underground Dust Conditions in Large South African Gold Mines,* Pneumoconiosis Research Unit Report No. 3/63 (Johannesburg: South African Council for Scientific and Industrial Research, 1963): 13, 22.

[58] D. G. Beadle, 'An Epidemiological Study of the Relationship between the Amount of Dust Breathed and the Incidence of Silicosis in South African Gold Miners' *Proceedings of the Mine Medical Officers' Association* 45, no. 391 (1965): 31–37.

histories were correlated, the results indicated a clear relationship between the rate of silicosis and total shifts worked. There was also a rise in the incidence of tuberculosis over time. Beadle concluded that diagnosis based on X-rays underestimated the incidence of occupational disease.

Beadle's most important findings were on dust exposures. His data suggested that prolonged exposure at existing levels was sufficient to cause silicosis and that the dust levels had not improved over a prolonged period: 'It is known that no dramatic changes in average dust conditions in South African gold mines have occurred since 1934' he wrote. 'Although some sources of dust production have been reduced, other mining processes which produce dust have been introduced or widely extended.'[59]

The research by Becklake, Chatgidakis, Laing and Beadle threw into question the value of medical surveillance and the Chamber's claims about mine safety. There was more dust and more active tuberculosis than was being compensated by the Bureau. Read together, the studies suggest that the industry was in crisis. Surprisingly, the pneumoconiosis conferences held in Johannesburg in 1959 and 1969 made no reference to those problems.

The 1959 conference was convened by the South African Institute of Medical Research at the request of the South African government to review the current literature and identify a research agenda. Experts from fifteen countries attended and the WHO sent a representative. The head of the Chamber, P. H. Anderson, welcomed the overseas visitors. Sarel Oosthuizen was the Conference President.

Eighty papers were presented.[60] Beadle spoke on methodology rather than the results of his important research on dust exposures, which was still in progress. Oosthuizen gave a paper on the early diagnosis of silicosis but said nothing about tuberculosis, which had been the focus of his Committee's report just four years earlier. Chatgidakis spoke about white miners but made no reference to her path-breaking research on black miners. The only paper on tuberculosis was by the British researcher A. Meiklejohn. The proceedings were dominated by J. C. Wagner's presentation on asbestos and mesothelioma in the Northern Cape.[61] Wagner's work changed the research agenda and the next two conferences, in Sydney in 1968 and Johannesburg in the following year, focussed upon asbestos disease.

The Pneumoconiosis Conference of 1969 was funded by the Chamber,

[59] Beadle, 'An Epidemiological Study': 33; see also D. G. Beadle 'Recent Progress in Dust Control in South African Gold Mines,' in H. A. Shapiro (ed.), *Proceedings of the International Conference on Pneumoconiosis, Johannesburg, 1969* (London and Cape Town: Oxford University Press, 1970): 69–78.

[60] A. J. Orenstein (ed.), *Proceedings of the Pneumoconiosis Conference held at the University of the Witwatersrand, Johannesburg, 9-24 February 1959* (London: Churchill, 1960).

[61] See Jock McCulloch, 'Saving the Asbestos Industry, 1960 to 2003,' *Public Health Reports* 121 (2006): 609–14.

the Department of Mines and the asbestos industry. Oosthuizen was again elected President. In the decade since the previous Johannesburg meeting there had been an explosion of asbestos research, and in 1969 around 40 per cent of the papers were devoted to the 'magic mineral'.[62] Beadle's two papers on silica dust were simply lost. [63] There was no presentation on tuberculosis. The shift in focus from silicosis to asbestos proved beneficial for the mining houses. Asbestos accounted for less than 5 per cent of the value of South Africa's mineral exports and employed just over twenty thousand men and women. By 1965 the evidence of the asbestos hazard was overwhelming but the asbestos mines, some of which fell within the Anglo American group, resisted regulation and it took a further thirty years before the last of the mines in the Northern Cape was closed.[64]

The health of miners

By 1970 Oluf Martiny had been working at the WNLA hospital for fifteen years. In August of that year he gave a paper to the MMOA on the management of tuberculosis. It was a notable presentation as it was unusual for a medical officer to speak so candidly at such a forum. Martiny told his audience that the high incidence of infection in the rural areas was due to economic hardship. Most blacks lived in a remittance economy, but even when combined with wages from migrant work agricultural output could not provide a living. Malnutrition was present at all times and became catastrophic during dry seasons. Poverty explained both the susceptibility to tuberculosis and the reluctance of miners to undergo treatment. When Martiny told a miner he needed hospitalisation, the reply was often: 'When I am dead, I am dead, while I am alive I must work.' Treatment was rarely completed, and 95 per cent of those discharged from mine hospitals did not complete their treatment elsewhere. That failure added to the infective pool and created drug-resistant strains of the disease. 'The danger of tuberculosis today,' Martiny said, 'is not the tuberculotic, but the employer who casts him out to become a spreader of the disease.'[65]

Martiny estimated that fewer than 20 per cent of tuberculosis cases at the WNLA hospital received the compensation to which they were entitled. That in part explained why so few miners continued with treat-

---

[62] See H. A. Shapiro (ed.), *Pneumoconiosis: Proceedings of the International Conference, Johannesburg 1969* (Cape Town: Oxford University Press, 1970).

[63] At the Australasian Pneumoconiosis Conference in Sydney in 1968 Derrick Beadle gave a paper on silicosis and dust counts, but it was lost amidst the presentations on asbestos. There were no papers on the synergy between silicosis and tuberculosis. *First Australasian Pneumoconiosis: A Conference on Airborne Dust in Industry, Its Measurement, Control, and Effects on Health, Proceedings University of Sydney 12–14th February 1968.*

[64] See McCulloch, *Asbestos Blues*: 201–7.

[65] O. Martiny 'A Tuberculosis Programme for South Africa, the Mines and Voluntary Agencies' *Proceedings of the Mine Medical Officers' Association* 49, no. 407 (1970): 168, 162, 166.

ment.[66] He wanted payment of compensation or a pension provided on condition that miners remained under care. 'It is a cruel test in industry if its only purpose is to discard the wrecks or possible wrecks and if it is not aimed at helping the victims to be cured, thereby enabling them to continue working.'[67] Martiny suffered no repercussions for his comments, which were published by the MMOA; the industry was so powerful that such criticism was irrelevant.[68]

The lives of the miners about whom Martiny wrote were full of danger and stress. Miners were routinely awakened en masse at 3 a.m. irrespective of the time they began their shifts. Some missed breakfast, as it was too early to eat. If they did not eat until their return to the hostel in the afternoon, twenty-four hours would have elapsed since their previous meal. As one miner commented during a survey in the 1970s, 'When we arrive late in the hostel we find that there is no food and have to spend the night without food. Recently we spent four days without food because of delays in cages.'[69] The food was often badly prepared, and when men were late coming off shift the dining hall was sometimes closed. In winter latecomers might have to shower in cold water, which in the high veldt was not just unpleasant but dangerous. Because workers were awakened so early, many were in bed by 7 p.m., which left little time for leisure. When the Leon Commission visited mine hostels in 1994 it was shocked by the conditions in which food was prepared and served. The kitchens and the dining rooms appeared not to have been inspected by a health officer for years.[70]

On average, miners spent ten and a half hours a day working and travelling. Long delays at the cages and locos and long distances to walk meant that some men spent more than three hours getting to their workplaces. The work environment was brutal. In addition to the constant threat of rock falls, miners had to deal with intense heat, dust, noise, darkness, confined spaces as well as harassment and assaults by white supervisors. Most migrant workers made every effort to maintain contact with their families. The average time spent travelling home was eight hours, at an average cost of R8.00 per journey, which in 1980 was a little under 10 per cent of a mineworker's monthly wage. The actual time spent at home may have been as little as twelve hours. It often took ten days for a message to be conveyed from a miner to his family through The Employment Bureau of Africa (Teba)'s field offices. A full cycle of communication would take three weeks, which added to the men's sense of isolation.[71]

---

[66] Martiny, 'A Tuberculosis Programme': 164.
[67] Martiny, 'A Tuberculosis Programme': 162.
[68] Interview with Martiny.
[69] C. M. Laburn and P. de Vries, 'Factors Which May Be Causing Stress among Black Migrant Workers, the Concomitant Behavioural Reactions, and their Dissipation' *Proceedings of the Mine Medical Officers' Association* 61, no. 430 (1982): 13.
[70] See *Report of the Commission of Inquiry into Safety and Health in the Mining Industry* (Pretoria: Department of Minerals and Energy Affairs, 1995): 58.
[71] Laburn and de Vries, 'Factors Which May Be Causing Stress': 7, 8.

In April 1980 Martiny published a second review of the repatriation system which suggested that little had changed. The individual mines still had different approaches to treatment but, whatever the approach, the lack of contact between mine doctors and local health services meant there was no coordination. During visits to hospitals in recruiting areas Martiny found that 70 per cent of tuberculosis patients had worked on the mines. He also found that 70 to 90 per cent of miners did not continue with treatment on their return home. Part of the reason lay in the poverty he had observed ten years earlier and part in the fact that, because of bureaucratic mistakes and staff shortages, only a third of beneficiaries actually received their compensation payments.[72]

The failure of treatment was closely tied to the inadequacy of mine wages. Those wages were set at such low level that any interruption to a miner's income was a catastrophe. Men continued to work as long as they could. In 1980, data for families in the Eastern Cape with incomes of less than R1,500 per year show that 66 to 71 per cent of total income came from mine wage remittances, 14 to 19 per cent from pensions, and 11 to 15 per cent from local employment. Only 2 per cent came from subsistence agriculture. A 1995 study of the dependents of Anglo American employees in the Lusikisiki District found that 93 per cent had no income other than mine wages.[73] The vulnerability of labour-sending communities and the human cost of repatriation must have been obvious both to employers and regulatory authorities.

One of the roles of the Pneumoconiosis Bureau was to perform examinations of migrant workers in the African territories. During 1966 just twenty-four examinations were done at the request of the Director: sixteen in Lesotho, three in Swaziland, four in Botswana and one in Malawi.[74] The Chamber of Mines showed the same lack of concern for the fate of South African miners. In 1974 the gold industry began a ten-year capital research and development programme with a budget of R100 million. The aim was to refocus research on the four critical problems in mine safety and occupational health, namely rock bursts, mine cooling, the mechanisation of stoping, and human resources.[75] Dust, silicosis and tuberculosis were not part of that initiative.

In the decades after 1952, vaccination and detection campaigns, the use of antibiotics and continuing improvements in living and working conditions all but eradicated tuberculosis in high-income countries. None of those benefits reached South Africa's gold miners. Ironically, rather than reducing the risks facing miners and their families, the introduction of effective treatment led to even worse abuses of labour.

---

[72] O. Martiny, 'Socio-Medical Problems in the Mining Industry in Relation to Altered Recruiting' *Proceedings of the Mine Medical Officers' Association* 58, no. 427 (1980): 8, 9.
[73] Trapido, 'The Burden of Occupational Lung Disease': 68.
[74] *Report of the Miners' Medical Bureau for the Period 1st April, 1965 to 31st March, 1966* (Pretoria: Government Printer, 1966): 12.
[75] See *Report of the Commission of Inquiry into Safety and Health*: 93.

9
The Sick
Shall Work

The development of the Free State mines shifted investment and labour away from the Rand. By the early 1980s the new fields were employing two hundred thousand men. Most of the mines were operated by Anglo American, which built the nine-hundred-bed Ernest Oppenheimer Hospital at Welkom as its central facility. The largest and best-equipped hospital of its kind in Southern Africa, Ernest Oppenheimer was to become the site of the most influential medical research on gold miners. A second change during that period was the formation of the National Union of Mineworkers (NUM). Founded in 1982, the NUM was successful in gaining wage rises and soon had more than three hundred thousand members. The NUM was committed to improving mine safety and was instrumental in the establishment of the Commission of Inquiry into Safety and Health in the Mining Industry, usually known as the Leon Commission, which tabled its findings in 1995.

In the period leading up to majority rule, compensation payments under the Occupational Diseases in Mines Works Act No. 78 of 1973 (ODMWA) were based on race. Like its predecessors such as the Silicosis Medical Bureau, the Medical Bureau for Occupational Diseases (MBOD) delegated the examinations of black miners to mine medical officers. Over time the differences between the technologies used with white and black miners widened. Whites were given full-sized chest X-rays while miniature X-rays were used for blacks. The lack of pulmonary-function equipment at mine and rural hospitals contributed to the underreporting of chronic obstructive pulmonary disease among blacks.[1] The MBOD provided follow-up services for discharged and retired white miners, but there was no parallel service for blacks. In reviewing the system the Leon Commission commented: 'Practically, no facilities exist for the examination and investigation of former mineworkers, either in this country or adjoining states.'[2]

---

[1] Trapido, 'The Burden of Occupational Lung Disease': 8.
[2] *Report of the Commission of Inquiry into Safety and Health in the Mining Industry, Johannesburg: Government Printer, 1995*: 42.

Majority rule did not resolve that problem, and miners still struggle to receive compensation.

Because tuberculosis is a relapsing disease, under a medical orthodoxy stretching back to the 1830s, infected men and women have been discouraged from continuing in dust work. From 1916 the employment underground of men with tuberculosis was prohibited in South Africa under the Miners' Phthisis Acts. The industry's policy of repatriating infected men to rural areas where medical care was minimal contravened no law but it did run counter to accepted public health practice. Despite protests from the Stratford Commission and the Oosthuizen Committee, it was not until the 1980s that employers moved away from repatriation to the provision of treatment.[3] The context for that change was a restructuring of the industry.

From the early 1980s the mines pursued a policy of labour stabilisation. Migrant labour was modernised, but in ways that externalised the social, economic and health costs. Experienced miners were given the right of return to their previous jobs provided they did so within a specified time. The policy was made more appealing to labour by rising unemployment, the impact of drought and the tightening of influx controls. In 1976 the proportion of black miners who had worked for more than ten years was around 14 per cent: by 1990 it had risen to 37 per cent.[4] There was also an increase in the length of each contract, the average age of miners and the total period spent in service. Prior to stabilisation workers had commonly tended to leave the industry before silicosis or tuberculosis became manifest.[5]

The NUM agreed with management that restructuring was necessary and proposed the creation of a more skilled workforce. The mines chose to do the opposite, with closures, retrenchments and the introduction of subcontracting. Subcontracting, which took various forms from labour-only contracting to outsourcing specific jobs, was consistent with the mines' history of minimising labour costs.[6] The major area for subcontracting was underground. Much of the labour came from retrenched miners, who were insecure and non-unionised and who did the same work as before, often at the same mines, without the benefits they had previously enjoyed. Subcontractors undercut wage agreements and ignored occupational health and safety conventions, and they had a negative impact on the NUM's membership. In 1987 the mines employed around five hundred thousand men; ten years later they employed half that number. The effect of those job losses on South Africa's neighbours

---

[3] Roberts, *The Hidden Epidemic*: 49.

[4] Wilson, evidence before the Commission of Inquiry into Safety and Health in the Mining Industry, at Braamfontein, Johannesburg, 1 August 1994, South African National Archives: 734–36.

[5] See *Report of the Commission*: 47.

[6] See Jonathan Crush, Theresa Ulicki, Teke Tseane and Elizabeth Van Veuren, 'Undermining Labour: The Rise of Sub-contracting in South African Gold Mines' *Journal of Southern African Studies* 27, no. 1 (2001): 5–31.

can be gauged by their dependence on repatriated wages. Despite the sharp fall in employment in 1993, remittances to Lesotho amounted to R284 million or 55 per cent of the country's GDP. The estimated contribution to Swaziland's GDP was 20 per cent and to Botswana's 10 per cent.[7]

Reactivation

The research conducted by the SAIMR and the Chamber of Mines tended to justify existing practice rather than change work regimes. The exception was a research project carried out by Dr Robert Cowie at the Ernest Oppenheimer Hospital from 1977. On the basis of a single study, Cowie revolutionised the treatment of tuberculosis in South Africa's most important industry. By 1991 the policy of reactivation that he initiated had been adopted throughout the gold mines.

The aims of Cowie's project were outlined in presentations he and his colleague Dr B. C. Escreet gave to a meeting of the Mine Medical Officers' Association in 1978. Cowie reminded his audience that the system of treating tuberculosis had been a failure. Fewer than 10 per cent of miners completed the two-year-long regimen, and as a result many developed drug-resistant disease. In contrast, he argued, 95 per cent of men who remained in risk work while being treated could be expected to complete a six- or nine-month course of chemotherapy. The new drugs were expensive, but there would be major savings through increased productivity. Cowie believed that once his approach to treatment was adopted: 'The whole country will become a low prevalence community. Once that stage is reached there will no longer be need for six-monthly periodical X-rays … Health education and good nutrition will be sufficient to hold tuberculosis at bay.'[8] Escreet was equally confident that in the long term reactivation would see such a drastic reduction in chronic and drug-resistant cases that it would more than outweigh the cost. The rapid cure of tuberculosis would result in a lower prevalence and ultimately in better utilisation of labour: 'The benefits, both within the industry and the country, are inestimable … and, are likely to be profound.'[9] Reactivation was devised for black miners, who accounted for virtually all tuberculosis cases.

In 1987, Cowie submitted a thesis on silicosis among gold miners to the University of Cape Town in which he provides a history of reactivation. In so doing he drew a sharp distinction between silicosis and tuber-

[7] Submission of the Chamber of Mines to the Commission of Inquiry into Safety and Health in the Mining Industry, 1994, South African National Archives: 24.

[8] R. L. Cowie, 'The Modern Treatment of Tuberculosis' *Proceedings of the Transvaal Mine Medical Officers' Association* 57, no. 425 (1978): 56, 57.

[9] Dr B. C. Escreet, 'The Ernest Oppenheimer Hospital Tuberculosis Project: A Clinico-epidemiologic Study' *Proceedings of the Transvaal Mine Medical Officers' Association* 57, no. 425 (1978): 58.

culosis and, surprisingly, offered no comment on the synergy between the diseases. Silicosis he identified as such a trivial disease that men with dusted lungs could continue in risk work without endangering their health.[10] Cowie explained that an 'unstructured investigation' he conducted in the late 1970s, in response to requests from men with silicosis who wanted to continue in risk work, had shown that in the absence of tuberculosis there was no significant pulmonary dysfunction. Consequently:

> it was resolved that until a formal study could be made, men, who had been told of their disorder but wished to continue working as miners, would not be reported [to the director of the MBOD as required by law] as silicotic unless they were planning to leave the industry, had lung dysfunction, or were, on the basis of their young age, judged to be at risk of developing severe disease in the course of their potential period of future employment ... At the start of the present study [presumably 1979] the routine reporting of men with silicosis had been discontinued for approximately 5 years on the mines served by the Ernest Oppenheimer Hospital.[11]

The legality of that decision is unclear.

According to Cowie black miners had high dust exposures but a very low silicosis rate. In addition to silica dust, individual susceptibility and smoking were also contributing factors to pulmonary disability. Cowie conceded that miners with silicosis showed an increased prevalence of breathlessness but claimed it was not debilitating.[12] That profile of silicosis formed the foundation of Cowie's work on tuberculosis.

The tuberculosis project offering short-course treatment for black miners in risk work ran from April 1977 until December 1986. To make that possible, key sections of the mines legislation were suspended by the director of the MBOD, Dr F. J. Wiles. Cowie acknowledges that such a decision was unprecedented. 'Interestingly there was considerable concern that an outcry would occur if it became known that migrant gold miners who had been treated for pulmonary tuberculosis were being allowed to return to work. In the event, the return of these men to work was effected without publicity and without event.'[13]

The effectiveness of treatment was carefully monitored, and the men were reviewed with chest X-rays for up to five years after initial diagnosis. The measure of success was the relapse rate, which was defined as the recurrence of positive sputum (as we have seen, a notoriously unreliable

---

[10] R. L. Cowie, 'Silicosis, Pulmonary Dysfunction, and Respiratory Symptoms in South African Gold Miners' M.D. thesis, Faculty of Medicine, University of Cape Town, 1987: 92.

[11] Cowie, 'Silicosis': 11, 12.

[12] Cowie, 'Silicosis': 92.

[13] R. L. Cowie, 'Pulmonary Tuberculosis in South African Gold Miners: Determinants of Relapse after Treatment', Master's thesis, Department of Epidemiology and Biostatistics, McGill University, 1987: 10.

tool). According to Cowie's calculations, relapse occurred in 475 men or 13 per cent of the cohort. He concluded that previous exposure to silica dust, silicosis or continued exposure to dust after diagnosis were not the determinants of relapse; rather, the major risk factors were a miner's age and his home environment. On the basis of those results, in 1984 the regulations were altered to allow men who were under treatment with an approved regimen to return underground.[14] Reactivation was the cheapest way to treat tuberculosis in miners. It was also consistent with the medical philosophy that Schepers had seen in operation at the WNLA in the 1950s, namely, to get as much work as possible out of men before they broke down.

A critique

The results of the Ernest Oppenheimer Hospital Tuberculosis Project are open to dispute on a number of grounds. Cowie's most important data set is found in his Appendix Table 1A. The key to that set is the number of men classified as 'withdrawn': by the end of the first six months 425 of the original 1,085 had left the programme and presumably left the mines.[15] Some of those men may have died, others may have relapsed. After twelve months a further 303 had withdrawn, so that by the end of the first year 70 per cent of the cohort had vanished. At the end of the study period of thirty-six months only 58 men were left, and of those 10 had relapsed. The number of withdrawals undermined Cowie's finding that there was no variation in outcomes between the men who returned underground and those who did not. Even if the data set had been complete, the study ended too soon to detect the cumulative effects of prolonged exposure to dust.

In addition to Cowie's interpretation of his data, there were two other gaps in the project. Cowie admitted that there were no records of past dust exposures for his cohort, and he made little effort to fill that void. To collect such data would have required a detailed interview, which was conducted for only a subgroup of his study population. Like all black miners, Cowie's subjects had no medical histories. According to Cowie, prior to 1979 recruits 'would be given new documents, a new company identity number, and new medical history record for each contract. There was no official distinction between a man who had spent a life-time serving the industry and a man who was about to start his first contract.'[16] Cowie points out: 'As a result of this attitude, migrant workers in the South African gold mining industry were considered only in the context of their current contract.'[17]

Cowie's research and the reactivation policy that followed were based

[14] Cowie, 'Pulmonary Tuberculosis': 82.
[15] See Appendix Table 1A in Cowie, 'Pulmonary Tuberculosis': 113.
[16] Cowie, 'Silicosis': 11.
[17] Cowie, 'Pulmonary Tuberculosis': 4–5.

on the assumption that tuberculosis was not caused by dust in the mines. He also assumed that 'the apparently adequate management of tuberculosis disease in black South African gold miners has had no influence on the prevalence of disease because of the large infective pool that exists in recruitment areas [and therefore] the effective surveillance and treatment of the disease on the mines can never significantly influence the incidence of pulmonary tuberculosis'.[18] Cowie's views were not shared by Martiny, who during visits to hospitals in recruiting areas in the 1970s, found that 70 per cent of tuberculosis patients had worked on the mines.[19] They are also at odds with the findings of the 1943 Stratford Commission, of Schepers in 1952, of the Oosthuizen Committee in 1954 and of the recent research by the National Institute of Occupational Health.

The legal status of the project was ambiguous. Cowie writes: 'Dr Frank Wiles, the previous Director of the Medical Bureau of Occupational Diseases, made sure that the programme remained within the law by changing the laws.'[20] According to the Bureau's Annual Report for 1979, in that year the act had been amended to enable white and coloured miners who had been certified for tuberculosis and 'adequately treated to return to full-time work in risk areas'. Previously, the Report continued, they had been allowed a maximum of a hundred hours per month in a risk area, but because that restriction had caused undue financial hardship the ruling had been changed. According to the Director: 'If tuberculosis is discovered at an early stage modern treatment makes it safe for the patient to return to his normal work within 2 or 3 months.' That amendment did not, however, apply to black miners. If the results of a trial then under way at the Ernest Oppenheimer Hospital were satisfactory, the Report went on to say, it would be possible to recommend that black miners treated for tuberculosis and 'completely recovered' be allowed to return to full-time duty in risk areas.[21] I have found no other reference to a trial study of white and coloured miners, and the Director's use of the terms 'adequately treated' and 'completely recovered' as synonyms is confusing. The black miners in Cowie's study were not 'completely recovered': they were in treatment for a life-threatening disease. I have not been able to find a contemporary review of Cowie's research reports, and there are no peer reviews in the medical literature of the reactivation policy it inspired. Neither the industry nor the state attempted to replicate Cowie's results, and until 2005 there were no follow-up studies on the consequences of reactivation.[22]

It is difficult to determine how quickly the industry adopted reactiva-

---

[18] R. L. Cowie, M. E. Langton and M. R. Becklake, 'Pulmonary Tuberculosis in South African Gold Miners' *American Review of Respiratory Diseases*, no. 139 (1989): 1087.

[19] Martiny, 'Socio-Medical Problems in the Mining Industry': 8.

[20] Cowie, 'Pulmonary Tuberculosis': v.

[21] *Report of the Medical Bureau for Occupational Diseases, Republic of South Africa, for the period 1 April 1979 to 31 March 1980*. (Johannesburg: Government Printer, 1981): 5.

[22] Girdler-Brown et al., 'The Burden of Silicosis': 640–47.

tion. During the early 1980s the reports of the Compensation Commissioner for Occupational Diseases show that over four thousand black miners were compensated annually for tuberculosis but there is no data on treatment.[23] It is certain that after 1985 medical reactivation was gradually adopted as employers recognised the economic benefits of the new policy. Over time it became normalised. The Safety in Mines Research Advisory Committee's *Handbook on Occupational Health*, published in 2001, contains the following statement:

> Patients [with tuberculosis] may return to work if clinically fit to do so. These individuals are uninfectious and would not pose a threat to other employees. Cure rate for miners who return to work while taking TB treatment are good. The WHO supports the return to work of TB patients as soon as possible. This also applies to underground mining.[24]

The *Handbook* does not acknowledge the cruel choice that miners face between returning to risk work and losing their jobs. What constituted clinical fitness is not defined, and no data on cure rates are presented. How, for example, did those rates compare with those for men who did not return underground? The only evidence cited in the *Handbook* comes from the data sets Cowie completed two decades earlier.

The Leon Commission

While the industry was reaping the benefits of reactivation and the nation was heading toward majority rule, the Leon Commission was hearing evidence about mine safety and occupational disease. The Commission, which tabled its final report in 1995, was the first inquiry into the mines for thirty years, and it was also the first to hear testimony from black miners. Like all previous commissions, Leon acknowledged the economic importance of the gold industry, which provided a livelihood for as many as 2.8 million South Africans.[25]

The Chamber's submission to Leon argued that the industry's systems of medical surveillance, diagnostics, treatment and review were generally superior to the facilities available to the South African public. Furthermore, the legislative framework for the management, certification and compensation of occupational disease was more than adequate. Its concluding statements contained almost nothing about lung disease or migrant labour. While there was a section on tuberculosis, the issue was conflated with the problem of HIV/AIDS, which tended to pathologise the

---

[23] *Report of the Compensation Commissioner for Occupational Diseases 1990.* (Johannesburg: Government Printer, 1991): 9.

[24] G. J. Churchyard, 'Tuberculosis and Associated Diseases' in R. Guild, R. Erlich, J. R. Johnson and M.H. Ross (eds), *A Handbook on Occupational Health Practice in the South African Mining Industry* (Johannesburg: Safety in Mines Research Advisory Committee, 2001): 170.

[25] *Report of the Commission*: 29.

disease in terms of sexual conduct. There were whole sections on rock bursts and accidents but perhaps three references to dust and none to silicosis. The Chamber proposed a system of self-regulation under the control of mine management as the best means to improve work safety and health.[26]

Leon rejected outright the idea of a self-regulated industry and found no basis for the Chamber's claims that the systems of health and safety were adequate. On the contrary, it said, a comparative study of fatalities in mining covering nineteen countries for the period 1989 to 1991 had classified South Africa's mines as amongst the most dangerous in the world. Leon pointed to an industry-wide failure to collect data and a disturbing decline in the quality of both the statutory reports and the publications of the mining houses. Annual medical reports had been phased out by the Anglo American Corporation in 1983 and by Rand Mines in 1986.[27]

In particular, Leon considered the industry's failure to control tuberculosis a matter for grave concern, especially in the light of the spread of HIV. Although there had been some improvement, the hostels were still crowded and spreading infection. Leon was particularly critical of the industry's reliance on mass miniature radiography for diagnosis, given the numerous studies showing that the technique was ineffective. Citing the work of Beadle, it concluded that dust levels had probably been unchanged for fifty years and that there was no evidence of a decline in the prevalence or severity of occupational disease.[28]

Leon recommended that the right to information and the right to refuse dangerous work be included in the legislation and that the Department of Health work with the health departments of Lesotho, Mozambique, Malawi, Swaziland and Botswana to provide care for former miners. The Mine Health and Safety Act of 1996 created a tripartite health body on the basis of those recommendations. It made X-ray examinations compulsory on departure from the mines, an initiative that had been strongly recommended by the Oosthuizen Committee forty years earlier.

A view from the mines

The Leon Commission covered a large amount of ground, but it did not specifically review the reactivation policy. An account of how the policy played out on the mines is found in the career of a medical officer who worked at Harmony in the Free State during the 1980s and in testimony from miners themselves.

Rhett Kahn was born in Nigel, in what is now Gauteng, in January 1956. He spent his childhood in Johannesburg and studied medicine at the

---

[26] Submission of the Chamber of Mines: 49–50, 208.

[27] *Report of the Commission*: 38, 15–16, 50, 41, 16.

[28] *Report of the Commission*: 55, 13, 53, 54–55, 52.

University of the Witwatersrand, graduating in 1980. While waiting for army call-up he worked with Harmony Gold Mines (Rand Mines) at Virginia, in the Free State, and when he came out of the army in 1985 he rejoined the company. Kahn was the only physician at Harmony with an occupational health diploma and for a time was in charge of the chest diseases ward. There was a great deal of tuberculosis, but management's belief that tuberculosis was brought onto the mines from the rural areas went unchallenged. Medical officers were encouraged to get infected men out of the system as quickly as possible.[29]

Rhett Kahn had read Cowie's study but he was unconvinced by the supporting data. He conducted his own pilot survey at Harmony and found a relapse rate among those who returned underground of 7,596 per 100,000 compared with 1,284 among those who did not.[30] The figures were disturbing, and Kahn raised his concerns with management. Against the advice of Harmony's chief medical officer, in August 1991 he wrote to the Director of the MBOD, Dr P. Lombard. He reminded the Director that reactivation ran counter to a century-old medical ortho-doxy. It also ran counter to an agreement between the NUM and the Chamber that an employee who contracted tuberculosis be given surface work at 75 per cent of his pre-incapacity wage for six months while being treated. When, he asked, had the policy been changed? There was also the issue of Section 38(2)a of the ODMWA which stipulated that the senior medical officer at each mine hospital alone had the authority to decide the fitness of a miner to return to work. Had there been consul-tation between mine medical officers, mine management and the trade unions over reactivation? Finally, were miners being informed of the risks they faced in returning underground?[31] At Harmony, if a miner with tuberculosis was unwilling to return to work and there was no surface job available, he was retrenched. The Director did not reply to Kahn's letter. In January 1992 Rhett Kahn was fired. He filed a suit against Harmony for unfair dismissal and received a settlement. After leaving the company he worked part-time for the NUM and appeared as an expert witness before the Leon Commission. In 1999 he opened a private practice at Welkom.

Speaking of the introduction of reactivation at Harmony, Kahn told me that, rather than announcing a new policy and then implementing change, changes were made and then became conventional wisdom. Reactivation began as an experiment and was never acknowledged by Harmony or the other mining houses as a major innovation. The official tuberculosis rate was around 1 per cent, which for the industry as a whole translated into more than four thousand cases a year. It was costing the mines a lot of money. When Kahn was at Harmony, the choice was between firing

[29] Dr Rhett Kahn, interview at Virginia, Free State, 6 March 2010.
[30] Letter from Dr Rhett Kahn, Harmony Mine Hospital, Virginia, to Dr P. Lombard, Director, MBOD, Johannesburg, 13 August 1991, in author's possession.
[31] Letter from Dr Rhett Kahn.

infected miners or putting them back underground.[32]

Retrenchments in the early 1990s saw around two thousand men lose their jobs at Harmony. When a man was fired he was put onto a bus and got off the mine as quickly as possible to avoid protests. Former miners began coming to Kahn's surgery asking for help in applying for compensation. Those men had received no assistance from the union, and Kahn soon became known as the person who could process claims. Welkom was the catchment area for the Transkei, Lesotho and Mozambique, where there was a generation of men who were chronically ill. Kahn is still the only non-mine medical officer at Welkom dealing with those patients. Initially he was submitting around five hundred applications a year; today it is close to two hundred.

Much of Kahn's work is in treating the symptoms of silicosis with cortisone tablets and oxygen therapy. He also finds palliative care with asthma pumps useful. Although Section 36a(I) of the ODMWA refers to the right to treatment, there are no facilities to ensure that treatment takes place. Furthermore, it is not clear whether the Compensation Commission for Occupational Diseases or the employer is responsible. A factory worker with silicosis can obtain treatment and the Compensation Commission in Pretoria will pay, but that is not the case if the patient is a miner. That anomaly arises from the inequalities between the ODMWA and the Compensation for Occupational Injuries and Diseases (COIDA) Act No. 130 of 1993.

The compensation maze

The compensation system is fragmented between the Department of Mineral Resources, which controls accidents and the Department of Health, which is responsible for diseases. That fragmentation is made more significant by the inequalities in compensation and medical care that distinguish the mines from other workplaces.[33] The ODMWA covers only mining, while other occupations fall under COIDA, which is administered by Rand Mutual. COIDA provides far better benefits. There is, for example, a pension system for factory workers who contract silicosis whereas miners receive lump sums. The current payments are R35,000 for first-stage silicosis and between R60,000 and R84,000 for second-stage, none of them sufficient to sustain a family for long.

The inequalities in the compensation system extend to the regulation of the workplace. Department of Labour inspectors will go to a factory and investigate complaints about hazardous conditions. The inspector of mines will investigate dust, noise and accidents on mines but not an employer's failure to submit a compensation claim. There is no monitoring

---

[32] Rhett Kahn, interview.
[33] Janet Kahn, interview at Virginia, Free State, 4 March 2010.

of the implementation of the ODMWA and there are no procedures for disciplining breaches of the act. The MBOD has no inspectorate and therefore cannot investigate infractions.

In applying for compensation miners face a bewildering maze. The Bureau assesses claims, which it then forwards to the Compensation Commission in Johannesburg. It shares a computer and a filing system with the Compensation Commission, and both organisations are part of the Department of Health. The Compensation Commission is responsible for imposing levies on the mines and for awarding compensation for lung disease. Former miners are supposed to receive a two-year benefit examination that includes a chest X-ray and a lung function test. Any miner who has been treated for occupational tuberculosis is entitled to a medical examination twelve months after completing treatment. It is almost impossible for former miners outside South Africa to access those medical services.[34]

Because of a lack of capacity, it takes two to three years for a compensation claim to work its way through the system. At one point the MBOD had no radiologist and therefore could not process X-rays. It also lacks staff to conduct hearings. Of the applications submitted in 2009, fewer than 17 per cent were processed in that year. That was an improvement over 2006, when of 5,325 claims submitted only 7 per cent were resolved.[35] Unemployment for miners and their families means debt that, by the time a lump sum payment arrives, can be overwhelming. The Bureau has not published an annual report since 1994.

The compensation system is heavily bureaucratic and understaffed, and its lack of capacity is compounded by the attitude of industry. Employers have a vested interest in slowing down the application process and paying out as few awards as possible. Many of the system's worst features are residues of earlier legislation that was designed to minimise benefits. Since the Bureau's office at Welkom was closed in 1996, most cases have been referred to mine medical officers.

Applicants face many hurdles including distance and language, but none is more challenging than collecting the required documentation.[36] To be certified, a miner needs an X-ray, a lung function test, a benefit examination performed by a qualified physician and a record of service. A man may work on several mines over a lifetime, and often the records are incomplete. He may have to pay if he wants a print-out of his service details. Applicants need a bank account and a set of fingerprints from the police to establish identify. Widows must produce a post-mortem report and medical records as proof of illness. The Compensation Commission has vernacular speakers, but the process is so complex and drawn out that many applicants simply give up.

---

[34] See *The Mining Sector*: 9.
[35] Press Release by Andrew Louw, the Democratic Alliance shadow minister of labour, 31 March 2010.
[36] Janet Kahn, interview.

The Permanent Disability Certificate was introduced in the mid-1990s following negotiations between the NUM and industry. With the certificate, a man can never again work on a mine but receives a lump sum based on his salary and the years he has served. If over fifty-five a miner can access his provident fund. He can also apply separately for compensation to the MBOD.

A miner with tuberculosis who is treated can return to work, while a man with second-stage silicosis is not permitted underground. In an economy in which there are high levels of unemployment, a miner may be better off in the short term to avoid diagnosis. There is no compensation for HIV/AIDS, and infected men will hide their health status so that they can remain at work. Their immune systems are compromised and if they stay underground they are at extreme risk of contracting tuberculosis. For such men mining is a death sentence.

There has been a great deal of compensation fraud, much of it involving identity theft. For that reason, since 2008 no payments have been made to miners from Mozambique. Miners are prey to agents or touts who will offer to guide them through the bureaucracy.[37] By law agents are allowed to charge around 2 per cent of the final payment, but some touts will demand up to 50 per cent. Kahn has a sign in his surgery in Xhosa and Sotho advising miners not to pay an agent. Janet Kahn recalls a woman named Gertrude who was bringing widows to the surgery, paying for the X-ray and the taxi and then demanding a large part of the settlement. Gertrude was prosecuted in the local magistrate's court and forced to refund money to her clients.

Miners have few assets to shield them from misfortune. Their families are dependent upon a male wage and the loss of work means debt from which there is no escape. They usually have more than a single episode of illness and often they suffer from more than one disease. They become ill, are treated, go back to work and then relapse. The miners I interviewed at Welkom in March 2010 want to work but cannot. To gain compensation is a struggle and often they are forced to rely upon the goodwill of their former employers.

Ntja Moletsane was born at Mohale's Hoek in Lesotho in January 1948 and began working at the West Rand mine in Gauteng in 1969. There was a lot of dust, he said, and no warnings about the risks. He then worked at the President Steyn mine in Welkom, where many men had *sifuba* or lung disease. In 2004 he was certified with first-stage silicosis and received R30,000 in compensation. In July 2005 he received a further R20,000 for second-stage silicosis. In contravention of the law, he continued to work underground until January 2006, when he was diagnosed with tuberculosis. After being treated he returned underground once more. Ntja Moletsane has six grown children. He has land and cattle in Lesotho but can no longer do physical work.[38]

[37] Janet Kahn, interview.
[38] Ntja Moletsane, interview at Welkom, Free State, 4 March 2010.

Zakaria Mofokeng Mokaeane was born at Leribe in Lesotho. His father was a miner who had *sifuba* but received no compensation. Zakaria started work at Rustenberg Platinum in 1969 and was a team leader and a member of the NUM at Western Holdings from 1970 to 1996. He then moved to the President Steyn mine at Welkom, where at the training centre he was told that the dust was dangerous. He worked in the stopes, where it was very dusty. There were no masks, so he tied a handkerchief over his face. He became ill in 2005 and was in hospital for three weeks, after which he went back underground for the next three years. He received R35,000 for first-degree silicosis and has applied for second-degree compensation. He has seven children, the youngest of whom is fifteen. He is chronically ill and says that he can't get air into his lungs.[39]

Zakaria's brother Khanhelo Mokaeane spent much of his life working at Western Holdings in Welkom. In 1973 he was recruited by Teba as a driller. At the training centre he was warned about the dust and the risk of tuberculosis and phthisis. The miners were instructed to water down and to wear masks. Many men he worked with became ill, and many died. He thought that he would eventually get *sifuba*. In 1993 he was diagnosed with tuberculosis. He spent a week in hospital and went back underground while being treated. He was certified in February 2008 with first-degree silicosis, and compensation was paid. He returned home to Lesotho, where he has eight children.[40]

Oscar Dlamini was born in Swaziland in November 1960. His father worked on the gold mines but did not contract *sifuba*. At the age of nineteen Oscar started at the Doorfontein mine in Johannesburg. He was a winch driver, and there was a lot of dust. After the blasting it was his job to check the work site. He was the first person to enter the chamber, when the dust was heaviest. He was given no warnings about the dangers. When he started at Harmony in 1986, he was warned about the dangers of dust during training sessions. There was visible dust at Harmony, and in 2001 he was diagnosed with silicosis. He continued to work until October 2009, when he became very ill. He was then medically repatriated. Many of his former workmates have *sifuba*. Oscar Dlamini has a massive fibrosis and only 38 per cent of lung function. His life expectancy is less than five years. Like other miners, he has struggled to obtain compensation and medical treatment. 'Those with tuberculosis at the mines,' he told me, 'receive treatment, but there is nothing for silicosis. The mine doctors are no good for silicosis.' He became a South African citizen in 1996. He has five children, the three youngest still at school. The family income is his unemployment insurance, which will last only eight months.[41]

Reactivation, which often sees a period of remission followed by relapses, puts family members at risk. Aline Tsoene was born at Hobhouse in the Eastern Cape in 1973 and has lived in Welkom since 1992. She has

[39] Zakaria Mofokeng Mokaeane, interview at Welkom, Free State, 4 March 2010.
[40] Khanhelo Mokaeane, interview at Welkom, Free State, 9 March 2010.
[41] Oscar Dlamini interview at Welkom, Free State, 9 March 2010.

two children, aged eleven and three. Her husband, Bennett Nkhotee, is a miner from Lesotho and is in his fifties. In 2001 he was diagnosed with tuberculosis, and he continued to work underground while being treated. He relapsed in 2006. In June 2009 he became ill again and the Ernest Oppenheimer Hospital diagnosed him with silicosis. He can no longer work. The mine helped with his application to the provident fund. Like many of the miners at Welkom, he is a Lesotho citizen while his children are South African. When Lesotho men lose their jobs they lose their place in the South African economy and in the health system.

When I interviewed Aline Tsoene at Virginia, her husband was in Lesotho looking after their cattle and their garden. The family was in financial crisis. There were no new clothes for the children, and Aline was buying less meat and milk. The rent was manageable, and she received a state grant of R240 a month for each child. Her youngest child, Rothabile, had been treated for tuberculosis that he had contracted from his father. (Nkhotee had not been advised to have his family screened.) The family's income was greatly reduced, and there was no prospect that the situation would improve. On the contrary, it was certain that the financial demands of her husband's illness would increase. Over time there would be taxi fares for treatment and the cost of medication. Aline Tsoene did not want to think about the future.[42]

All the miners I interviewed had spent more than a decade underground, but their knowledge of the risks of lung disease was limited and fragmented. That is consistent with a survey done by the National Institute for Occupational Health (NIOH) in 2007 to determine the levels of knowledge among managers, mine health officers and miners themselves. The survey revealed that silicosis (which miners called 'phthisis') along with tuberculosis and HIV/AIDS was little understood.[43] The report suggests that despite all the efforts by industry, the state and the trade unions since the Leon Commission, miners still have little awareness of risk.

Rhett Kahn's career at Welkom and the testimony of miners highlight the barriers to compensation which persist. It also shows that despite the political transformation since 1994 that the mines have continued to shift the costs of occupational disease onto rural communities and the state. That outcome has such deep historical roots it is hard to imagine it has been accidental.

The outcome

The best way to gauge the effectiveness of reactivation is through the rates of infection since the programme began. Since 1962, the incidence of tuberculosis in white miners has declined steadily while the official rate

---

[42] Aline Tsoene, interview at Virginia, Free State, 4 March 2010.
[43] See *A Summary Report of Research*.

for black miners has risen. Between 1965 and 1985 it increased from 3 to 10 per 1,000, and since 2000 researchers have consistently found high infection rates among former miners. Radiological evidence from Libode puts the incidence as high as 40 per cent.[44] The prevalence of tuberculosis amongst Basotho miners is 26 per cent.[45] Those results are supported by autopsy data from the NIOH, where the tuberculosis rate among black gold miners at autopsy rose from 171 per 1,000 in 1999 to 406 per 1,000 in 2007. The annual incidence of tuberculosis amongst miners is currently around 420: the rate for South Africa as a whole is 55.[46]

Since 1995 the WHO has promoted Directly Observed Therapy Short-course (DOTS) as the preferred treatment for tuberculosis. Since its adoption as the national strategy the incidence of tuberculosis in South Africa has almost doubled.[47] The obvious reason is HIV/AIDS. In 1987, 0.03 per cent of miners from areas other than Malawi tested HIV-positive; by 2000 the figure was 27 per cent.[48] Most Lesotho miners enrolled in treatment programmes for drug-resistant tuberculosis have been treated unsuccessfully a number of times, and the majority are HIV-positive. Even though South African gold mines, with their migrant workers and single-sex hostels, have been an ideal environment for sexually transmitted diseases, HIV/AIDS is not the only reason for the pandemic.

David Stuckler's recent research suggests that in sub-Saharan Africa higher rates of mining are associated with higher rates of tuberculosis, regardless of the impact of HIV/AIDS. His data show that a 10 per cent increase in mining corresponds to a 0.9 per cent rise in the overall tuberculosis rate. For the region as a whole, that meant seven hundred and sixty thousand excess cases in 2005. 'Miners in sub-Saharan Africa,' Stuckler writes, 'have a greater incidence of Tuberculosis than do any other working population in the world and constitute one of the largest pools of employed men in sub-Saharan Africa.' That relationship is most pronounced for gold mining. According to Stuckler: 'This suggests that gold mining is associated with a specific increased in risk of tuberculosis independent of HIV, which may plausibly be a result of the increased risk of silicosis.'[49] Mining also impacts on the mortality rates for the general community.

Jaine Roberts' 2009 study of former miners from the Ntabankulu District in the Eastern Cape confirms Stuckler's findings. Compensation for tuber-

[44] Trapido, 'The Burden of Occupational Lung Disease': 29, 143, 160.

[45] Girdler-Brown et al., 'The Burden of Silicosis': 640–47.

[46] Roberts, *The Hidden Epidemic*: 55.

[47] Imelda Bates, Caroline Fenton, Janet Gruber, David Lalloo, Antonieta Medina Lara, S Bertel Squire, Sally Theobald, Rachael Thomson and Rachel Tolhurst, 'Vulnerability to Malaria, Tuberculosis, and HIV/AIDS Infection and Disease, Part 1: Determinants Operating at Individual and Household Level' *Lancet Infectious Diseases*, no. 4 (2004): 268.

[48] David Rees, Jill Murray, Gill Nelson and Pam Sonnenberg, 'Oscillating Migration and the Epidemics of Silicosis, Tuberculosis, and HIV Infection in South African Gold Miners' *American Journal of Industrial Medicine* 53 (2010): 400.

[49] David Stuckler et al., 'Mining and Risk of Tuberculosis': 3, 1, 4, 179–95.

culosis is limited by law to in-service miners and those certified within twelve months of leaving the mines. After the lapse of a year, a claim can only be made when silicosis is also diagnosed. Over 70 per cent of the miners in Roberts's cohort of 205 men were diagnosed with tuberculosis more than twelve months after leaving the mines.

Much of the disease burden in the Ntabankulu District falls on young men. Close to 20 per cent of Roberts's sample were under fifty, and a further 27.3 per cent were between fifty and sixty years of age. Consequently, there was no state social security net for those men and their families, and they were living in poverty.[50] Respiratory illness was almost universal: 95.6 per cent were coughing, 71.2 per cent experienced dyspnoea or breathlessness, 82.4 per cent had fever, 80.9 per cent were in pain and 83.4 per cent had weight loss. The tuberculosis rates were also very high, and yet those diagnosed while in mine service made up only 26.3 per cent of the sample. Knowledge of the ODMWA regulations among former miners was virtually non-existent. Almost 90 per cent had received entrance medical examinations, but 85 per cent had not been given a medical on leaving the mines.[51]

In addition to spreading disease the migrant labour system is one of the major obstacles to medical care.[52] Treatment at the mines, in the public health systems of South Africa and in the neighbouring states such as Lesotho is uncoordinated. Some mines provide on-site care or refer miners to public health centres, while some return them to Lesotho for treatment. Each system captures a different set of patients and generates a different set of data. The restructuring of the labour force has created a further problem. Most men are employed directly by the mines, but those who are employed by private contractors are not entitled to the same health benefits. South Africa immigration law forbidding the employment of non-citizen novices on the mines has led to a growing pool of undocumented migrant workers.[53] There is an urgent need for the coordination of the health care between employers and the Departments of Home Affairs, Labour and Health in South Africa and Lesotho. There is also need for comprehensive education about HIV/AIDS, tuberculosis and silicosis and miners' rights to compensation.

[50] Roberts, *The Hidden Epidemic*: 54, 150.
[51] Roberts, *The Hidden Epidemic*: 152, 151, 79, 81.
[52] *The Mining Sector*: 4.
[53] *The Mining Sector*: 14.

10
Men
Without Qualities

The history of silicosis in South Africa is filled with paradoxes. The Rand mines were the first in the world to invest heavily in dust extraction technologies and instruments, such as the konimeter, to reduce risk. South Africa was the first state to recognise silicosis and tuberculosis as occupational diseases, and the gold mines were the first to use radiography to screen workers. Yet South Africa was unsuccessful in making the mines safe or in providing adequate compensation. The major paradox is between the intensity of public debate about silicosis and the invisibility of the disease burden. In the period from 1902 to 1990 there were twelve select committees, twelve commissions of inquiry, and four official and interdepartmental committees on miners' phthisis. The more the Department of Mines and the Chamber of Mines talked about silicosis and the more data they collected, the less the disease was visible. In South Africa as in the USA, the UK and Australia, the silicosis hazard has been pushed into the public domain by the efforts of organised labour, political activists, independent scientists and regulatory authorities. It was not made visible by capital.

South Africa was a special case, with its racialised state and labour markets dominated by a handful of corporations. The mines' importance to the national economy acted as a constraint on workplace reform. The gold mines were also quintessentially modern. There was medical monitoring to ensure the fitness of workers and compensation for those who became ill or injured. Oscillating migration commodified labour. As Robert Cowie remarked in 1987, 'The migrant workers had no individual qualities. A man with many years of service was considered equivalent to a man who was joining the industry for the first time.'[1] The gold mines externalised the costs of production to the labour-sending areas, while the compound system allowed employers to exercise high levels of control in the workplace. Finally, there was an affinity of interests between the state

[1] Cowie, 'Pulmonary Tuberculosis': 4.

and capital that guaranteed the mines' profitability. The system was stable, but there were points of fragility. The Chamber needed to placate the white trade unions and their political supporters. It also had to placate the imperial authorities in London, and the International Labour Organization with regard to the health of Tropical labour, and there were occasional fractures in the science such as the work of Andrew Watt and the post-mortem data from the Medical Bureau for Occupational Diseases (MBOD).

The election of the African National Congress government in 1994 brought a dramatic shift in the legal possibilities for the victims of occupational injury. The initial beneficiaries were former asbestos miners and their families. Asbestos mining by the British companies, Cape plc and Turner & Newall, was carried out in South Africa for almost a century. Working conditions in the Northern Cape and the Northern Province were hazardous, and when the mines closed in the mid-1980s they left behind large numbers of men and women with lung disease.[2] Extensive lobbying by community groups led eventually to a claim against Cape plc which began in a London court in February 1997. The plaintiffs were represented by the British lawyer Richard Meeran. Over the next three years the case was stalled over the question of jurisdiction. Cape wanted the case heard in South Africa, where the injuries had taken place and the plaintiffs lived. The plaintiffs wanted the case heard in the UK, where Cape's assets were held. After almost seven years of tortuous legal argument and appeals, the Law Lords finally ruled unanimously in favour of the plaintiffs. While the Cape case was running, a second set of asbestos claims was lodged in South Africa in August 2002 against the corporate giant Gencor. Those claims were made by the civil rights lawyer Richard Spoor on behalf of more than four thousand men and women who had worked for Gefco, a wholly-owned subsidiary of Gencor. It was the first time that such a claim had been brought in South Africa for injuries sustained in the mining industry.

In March 2003 Cape plc and Gencor reached out-of-court settlements in London and Johannesburg.[3] The settlements were notable for a number of reasons. They were class actions involving thousands of claimants. The Gencor case was the first time that South African miners had won a civil action for injuries in the workplace. The London plaintiffs had breached the corporate veil that Cape's lawyers had assumed would protect the company from its South African litigants. Documents spilled into the public domain revealing that the Departments of Mines and Health were aware for decades of the hazardous conditions on the asbestos mines but did nothing.[4] The collusion between state agencies and industry stretching

---

[2] See J. E. Roberts, 'What is the Price of 80 KGs? The Failure of the Detection of, and Compensation for, Asbestos-related Disease: Social Exclusion in Sekhukhuneland', Master's thesis, University of Natal, 2000.

[3] See Jock McCulloch, 'Beating the Odds: The Quest for Justice by South African Asbestos Mining Communities' *Review of African Political Economy* 32, no. 103 (2005): 63–77.

[4] Jock McCulloch, 'Asbestos, Lies, and the State: Occupational Disease and South African Science' *African Studies* 64 (2005): 201–16.

back to the 1920s was so profound that it called into question the integrity of the regulatory system as a whole. The day after the R450 million Gencor agreement was signed, Richard Spoor announced he would turn his attention to the plight of gold miners.

The stories of asbestos and gold mining are connected by the careers of the leading medical researchers in the UK, the USA and South Africa, who from the 1920s worked on both silicosis and asbestosis. They are connected by the US corporations, such as Johns Manville and Union Carbide that, over a period of decades, have faced litigation from employees suffering from those two diseases. As Richard Meeran has remarked: 'The chronology of the hazard is also very similar. In this case, the mines and the government had specific knowledge of the dangers of dust and the fact that it caused silicosis. While the asbestos issue was subjected to litigation for years (in the US and the UK) the gold mining industry has managed to escape justice.'[5] Of particular importance has been the role played by the mining company Charter Consolidated.

The Anglo American Corporation and its associate De Beers are vast enterprises. By 1957 they controlled 40 per cent of South Africa's gold production, 80 per cent of the world's diamonds and a sixth of its copper. They also produced most of South Africa's coal. From the 1930s Anglo American developed a maze of interlocking directorships, mutual agreements and restrictive trade practices. That complex structure, which is characteristic of the South African mining houses, makes identifying the ownership and control of subsidiaries difficult.[6]

Cape was a British firm with its head office in London, but from its foundation it was linked with De Beers and also to Anglo American. During World War II the Oppenheimer company Central Mining and Investment Corporation became the major shareholder in Cape. Central was one of four holding companies within the Oppenheimer group, the others being Anglo American, De Beers and Rand Selection. By 1949 Central Mining held the majority of seats on the Cape board. In 1969 the Central holding in Cape was superseded by Charter Consolidated, a British mining company, also controlled by Oppenheimer interests. Charter held 63 per cent of Cape's shares.[7] In 1979 Cape's mines were sold for £15.5 million to Barlow Rand, an Anglo American company. They were then sold to General Mining, yet another Oppenheimer enterprise. Charter's controlling share in Cape gave Anglo American a commercial interest in the asbestos industry. It also gave Anglo American's board a reason to monitor the Cape litigation. The final connection between the asbestos and the gold mine litigation is somewhat ironic. By the time the Cape plc

---

[5] 'Anglo Being Sued for R20m,' *The Sunday Independent,* 30 September 2007.
[6] See Duncan Innes, *Anglo-American and the Rise of Modern South Africa* (London: Monthly Review Press, 1984).
[7] See Geoffrey Tweedale and Laurie Flynn, 'Piercing the Corporate Veil: Cape Industries and Multinational Corporate Liability for a Toxic Hazard, 1950–2004,' *Enterprise and Society* 8 (2007): 7.

case reached court, the last of the asbestos mines in the Northern Cape had closed. Gold mining is today in its twilight, and its dominant role in the national economy has been over taken by the mining of manganese, chrome, coal and platinum.

Anglo American has almost limitless resources with which to defend itself, and much of the evidence about dust levels and disease rates lies in the corporation's hands. To date it has fought hard to have the Mankayi Mbini and Thembekile Mankayi cases (which, at the time of writing, are being fought in London and Johannesburg against AngloGold Ashanti) dismissed (see chapter 1). Three years after these suits by former miners with silicosis and or tuberculosis began, AngloGold Ashanti acknowledged the shortcomings of the compensation system. 'Many of these former employees may not have been diagnosed as suffering from the disease [silicosis] at the time they left the industry or later, in retirement, and they may not have received due compensation from the Compensation Commissioner.'[8] Despite that admission, there has been little improvement in the workplace. One of the best ways to gauge safety is through workers' knowledge of risk. In that regard a 2007 survey by the National Institute for Occupational Health (NIOH) presents a disturbing picture. Of a sample of 284 mine health and safety representatives, the majority were found to have little knowledge of the causes, prevention, or nature of occupational lung diseases. Only 17 per cent mentioned dust control as one of their responsibilities.[9] Almost two-thirds were unfamiliar with the term 'silicosis'.[10]

When Richard Spoor's case eventually reaches court, the defendants will no doubt cite various factors that contributed to the invisibility of disease. The industry's primary defence will be that the current high rates of silicosis and tuberculosis are a recent phenomenon that has arisen from the stabilisation of labour in the 1980s. While it is tempting to view that initiative as a break with the past, stabilisation has long been a feature of the gold mines. As early as 1906 Drs L. G. Irvine and D. Macaulay wrote that it was becoming common for East Coasters to remain for eighteen months continuously on the mines.[11] The Miners' Phthisis Medical Bureau's Annual Report for 1924 noted that retaining the services of experienced workers beyond the usual contract of six to nine months meant that a considerable number were permanent employees.[12] By 1930, of the four hundred thousand men who went through the system each year, 80 per cent were reengagements.[13]

[8] AngloGold Ashanti, *Report to Society 2006*, Johannesburg: 119.

[9] *A Summary Report of Research*, 5.

[10] *A Summary Report of Research*, 3.

[11] L. G. Irvine and D. Macaulay, 'The Life-history of the Native Mine Labourer in the Transvaal' *Journal of Hygiene* 6, no.2 (1906): 155.

[12] *Report of the Miners' Phthisis Medical Bureau for the Twelve Months Ending July 31, 1924*: 28.

[13] *Tuberculosis in South African Natives*: 74; see also *Report of the Commission on the Remuneration and Conditions*: 4–5.

Anglo American's second defence will be that silicosis is not easy to diagnose and its incidence was obscured by the lack of biomedical capacity in rural areas. While that excuse appears to have merit, on closer examination it tends to fall part. For almost a century Rand miners have been subject to medical review on entering and leaving the industry. In addition, from 1910 there have been skilled scientists in Johannesburg who could have monitored the post-employment health of migrant labour if only the Chamber had commissioned such research. At their peak the Rand mines employed over five hundred thousand men, which means that the impact of silicosis and tuberculosis must have contributed to the decline of rural economies within South Africa. The WNLA's failure to notify health authorities of repatriations meant that the impact was also felt in Nyasaland, Lesotho and Mozambique from the 1920s,[14] drawing protests from colonial administrations, and yet those practices continue to the present day.[15]

There have been a number of turning points in the mines' response to lung disease, but perhaps none has been more important than reactivation. The human costs of that policy, which were soon visible in a rising tide of tuberculosis, raise many questions. The most obvious is how it was possible for a major industry to put so many men and their families at risk. The subjects were black migrant workers who, under apartheid, had few political or civil rights and few choices with regard to employment. When Cowie reported that the Anglo American medical service at Welkom had numerous requests from silicotic miners to remain in risk work ('In general the men were obsessed with continuing their work, were unaware of and unconcerned about silicosis')[16] he failed to mention their choices: either to continue mining or to see their families destitute. When Cowie began his project, South African science was at its lowest ebb. According to Ian Webster and Tony Davies, who served as Directors of the NIOH from the 1960s to the late 1980s, the industry controlled the Institute. All research was submitted to the Chamber for approval before publication.[17] No researcher protested against the reactivation policy, if indeed they were aware that such a change had taken place.

Girdler-Brown and his colleagues' recent study of Lesotho confirms that the mines continue to promote infection in neighbouring states. Lesotho is a poor country and currently has one of the highest incidences of tuberculosis in the world.[18] Around 40 per cent of the adult male patients in Maseru's hospitals work or have worked in South African

---

[14] See the *Annual Medical and Sanitary Reports,* Zomba, Nyasaland, 1928 to 1940. Malawi National Archive.

[15] See *The Mining Sector.*

[16] Cowie, 'Silicosis': 11–12.

[17] Interviews with Ian Webster, 3 and 5 December 1997, and J. A. C. Davies, 26 and November 1998, National Institute for Occupational Health, Braamfontein, Johannesburg.

[18] Girdler-Brown et al., 'The Burden of Silicosis': 640–47.

mines, and 25 per cent of the drug-resistant cases in Lesotho are among former miners.[19]

The effectiveness of the WHO-approved Directly Observed Therapy Short-course (DOTS) treatment in developing countries has been compromised by the lack of health professionals. In addition the programme tends to shift attention away from the poverty, poor nutrition and gender inequalities that determine the incidence and severity of disease.[20] The incidence of tuberculosis is rising, and currently it is the leading cause of death from preventable infectious disease in the world. The WHO's Stop Tuberculosis target of halving the prevalence and fatality by 2015 is unlikely to be met.[21] South Africa and the surrounding states of Botswana, Lesotho and Swaziland, all of which have supplied labour to South Africa's gold mines, have among the world's highest rates of infection.

It is not possible to identify the moment when industry knew that the mines were spreading tuberculosis. There were, however, some obvious warnings, such as the 1912 Miners' Phthisis Commission and the papers from 1925 by Watt and Mavrogordato. Evidence of a serious risk was certainly available by the late 1920s, but the exact date is unknowable. It is possible, however, to identify who knew. Their ranks included the WNLA and NRC medical officers and management, who at first hand saw migrant workers with lung disease being repatriated without compensation; the staff at the MPMB and its successors the Silicosis Bureau; the senior figures within the Chamber, such as William Gemmill, and the members of the Gold Producers Committee (GPC), were aware of the scale of the problem; and there was the Johannesburg medical establishment, including Anthony Mavrogordato, L. G. Irvine and A. J. Orenstein, who controlled the science. The problem was almost certainly known to the senior officers in the Departments of Mines and Health, who helped draft the Miners' Phthisis Acts. Of course it was also known to anyone who read the post-mortem data published annually by the Miners' Phthisis Medical Bureau between 1925 and 1950. The spread of tuberculosis was an open secret that periodically erupted in the daily press, often in the form of exposés about the death trains.

As yet there has been no spill of documents into the public domain of the kind associated with asbestos and tobacco litigation in the USA. It is not yet known whether, like the senior executives of the US asbestos conglomerate Johns Manville, the GPC committed a set strategy to paper.[22] The lineaments of the system can be identified, however, in what was not done. They can be seen in the absence of follow-up medical surveys, the failure to carry out exit examinations, the systematic failure to notify colonial administrations of tuberculosis among repatriated workers and the

---

[19] *The Mining Sector*: 2.
[20] Bates, Fenton et al. 'Vulnerability to Malaria': 268.
[21] *Global Tuberculosis Control: A Short Update to the 2009 Report* (Geneva: World Health Organization, 2010): 26–28.
[22] McCulloch and Tweedale, *Defending the Indefensible*: 261–75.

lack of warnings to miners of the risks they faced. Tropicals were important to the mine's profitability. They were valued in part because they hid the cost. The human consequences were irrelevant.

Between the 1970s and 1994 the industry spent almost R1.5 billion on research and development to improve safety and productivity.[23] It is inexplicable that such investment failed to identify a pandemic that became visible with the advent of majority rule. Anna Trapido was a Ph.D. candidate with minimal material resources when she discovered a high rate of lung disease among former miners in the Eastern Cape, a problem that may well have existed for sixty years. The evidence suggests that the flaws in mine medicine were the result of industry policy. Those flaws were created and carefully maintained so that each medical innovation, be it the use of the stethoscope around 1912, the introduction of mass miniature radiography in the early 1950s or the reactivation policy since 1980, allowed employers to avoid their responsibilities.

The failures of the South African system of medical surveillance, data collection and compensation underpinned the commercial success of the nation's most important industry. The costs of production were shifted to rural communities within and outside South Africa's borders, and it is only now, more than a century after the mines began production, that those costs may at last be called in.

---

[23] Submission of the Chamber of Mines: 9.

SELECT BIBLIOGRAPHY

Unless otherwise indicated all archival materials cited are from the National Archives of South Africa, Pretoria. Materials from any other archival source are identified individually. In 1977 the Native Recruiting Corporation (NRC) and the Witwatersrand Native Labour Association (WNLA) were amalgamated to form The Employment Bureau of Africa or Teba. The Teba Archive is held at the University of Johannesburg.

Reports

Report of the Miners' Phthisis Commission 1902–1903 (Weldon Commission). Pretoria: Government Printer, 1903.

Report to the Secretary of State for the Home Department on the Health of Cornish Miners (Haldane Commission). London: H.M. Stationary Office, 1904.

Report of a Commission into Miners' Phthisis and Pulmonary Tuberculosis (Medical Commission). Cape Town: Government Printer, 1912.

Gorgas, W.C. *Recommendation as to Sanitation Concerning Employees of the Mines on the Rand made to the Transvaal Chamber of Mines.* Johannesburg: 1914.

Report of Select Committee on the Working of the Miners' Phthisis Acts Parliamentary SC 10-15. Third Report AN 4923, 1915.

Third and Final Report of the Select Committee on the Working of the Miners' Phthisis Acts. Union of South Africa Select Committee 10-15. AN 492. April 1916.

Report of the Technical Commission of Inquiry to Investigate Miners Phthisis and Pneumoconiosis in Metalliferous Miners at Broken Hill. New South Wales Department of Labour and Industry, Sydney: Government Printer, 1921.

Report of the Select Committee on Working of Miners' Phthisis Act, 1919. SC: 10-24, 1924 AN 451 Union of South Africa. Printed by Order of the

House of Assembly April, 1924.

Report of Tuberculosis Survey of the Union of South Africa by P. Allan. Cape Town: Cape Times Ltd, Government Printers, 1924.

Report of the Miners' Phthisis Commission of Enquiry 1929–30 (Young Commission). Union of South Africa. Pretoria: Government Printer, 1930.

Report of the Commission of Enquiry into the Position of Miners' Phthisis Beneficiaries. Pretoria: Government Printer, 1935.

Abraham, John C. *Report on Nyasaland Natives in the Union of South Africa and in Southern Rhodesia.* Zomba: Government Printer, 1937.

Report on Nyasaland Natives in the Union of South Africa and in Southern Rhodesia by J. C. Abraham. Zomba: Nyasaland Protectorate,1937.

The Prevention of Silicosis on the Mines of the Witwatersrand (Pretorius Commission). Pretoria: Government Printer, 1937.

Workmen's Compensation for Silicosis in the Union of South Africa, Great Britain and Germany. International Labour Office. Studies and Reports. Series F (Industrial Hygiene) No.16. London, 1937.

Report of the Commission Appointed to Enquire into the Possible Prevalence and Origin of Cases of Silicosis and Other Industrial Pneumonoconioses in the Industries of the Colony of Southern Rhodesia, and of Pulmonary Tuberculosis in Such Industries. Salisbury: Government Printer, 1938.

Report of the Miners' Phthisis Acts Commission, 1941–1943 (Stratford Commission). Pretoria: Government Printer, 1943.

Report of the Witwatersrand Mine Native Wages Commission on the Remuneration and Conditions of Employment of Natives on the Witwatersrand Gold Mines. Pretoria: Government Printer, 1943.

Report of the Commission on the Silicosis Legislation (Northern Rhodesia) Lusaka: Government Printer, 1949.

Report of Commission on Conditions of Employment in the Gold Mining Industry. Pretoria: Government Printer, 1950.

Report of the Commission of Enquiry regarding the Occurrence of Certain Diseases, other than Silicosis and Tuberculosis, Attributable to the Nature of Employment in and about Mines (Allan Commission). Pretoria: Government Printer, 1951.

Report of the Departmental Committee of Enquiry into the Relationship between Silicosis and Pulmonary Disability and the Relationship between Pneumoconiosis and Tuberculosis (Oosthuizen Committee). Pretoria: Government Printer, 1954.

Lambrechts, J. de V. *Investigation Undertaken by Anglo American Corporation of South Africa Limited an Industry Survey of Underground Dust Conditions in Large South African Gold Mines* South African Council for Scientific and Industrial Research Pneumoconiosis Research Unit Report No. 3/63 Johannesburg, 1963.

Laing, J. G. D *An Investigation into Tuberculosis in the Mining Industry,*

Transvaal And Orange Free State Chamber of Mines Research Organisation C.O.M. Reference: Project number 732/64A Research Report NO. 80/67, 1967.

Report of the Commission of Enquiry on Occupational Health (Erasmus Commission). Pretoria: Government Printer, 1976.

Report of the Commission of Enquiry into Compensation for Occupational Diseases in the Republic of South Africa (Nieuwenhuizen Commission). Pretoria: Government Printer, 1981.

Report of the Commission of Inquiry into Safety and Health in the Mining Industry (Leon Commission). Pretoria: Department of Minerals and Energy Affairs, 1995.

Arkles, R. S., Weston, A. J., Malekela, L. L. and Steinberg, M. H. *The Social Consequences of Industrial Accidents: Disabled Mine Workers in Lesotho.* National Centre for Occupational Health. NCOH Report No. 13, 1990.

A Summary Report of Research with Mine Health and Safety Representatives. What Does Silicosis Elimination Mean to Mine Health and Safety Representatives? SIM 030603 Track C Silicosis Elimination Awareness for Persons Affected by Mining Operations in South Africa Mine Health and Safety Council and National Institute of Health, 2007.

The Mining Sector: Tuberculosis and Migrant Labour in South Africa. Johannesburg: AIDS and Rights Alliance for Southern Africa, July 2008.

Published works

Alcock, Antony. *History of the International Labour Organisation.* London: Macmillan, 1971.

Allen, V. L. *The History of Black Mineworkers in South Africa.* Keighley: The Moor Press, 1992.

Ally, Russell. *Gold and Empire: The Bank of England and South Africa's Good Producers 1886–1926.* Johannesburg: Witwatersrand University Press, 1994.

Ashforth, A. *The Politics of Official Discourse in Twentieth Century South Africa.* London: Oxford University Press; Oxford: Clarendon, 1990.

Beinart, William. *Twentieth-Century South Africa.* Oxford: Oxford University Press, 1994.

Cartwright, A .P. *Doctors of the Mines: A History of the Work of Mine Medical Officers.* Cape Town: Purnell, 1971.

Castleman, Barry I. *Asbestos: Medical and Legal Aspects,* 5th edn. New York: Aspen Publishers, 2005.

Chanock, Martin. *The Making of the South African Legal Culture, 1902–1936: Fear, Favour and Prejudice* Cambridge: Cambridge University Press, 2001.

Cherniack, Martin. *The Hawk's Nest Incident: America's Worst Industrial Disaster.* New Haven: Yale University Press, 1986.

Davies, Robert H. *Capital, State and White Labour in South Africa 1900–1960: An Historical Materialist Analysis of Class Formation and Class Relations*. New Jersey NY: Humanities Press, 1979.

Dembe, Allard E. *Occupation and Disease: How Social Factors Affect the Conception of Work-related Disorders*. New Haven CT: Yale University Press, 1996.

Derickson, Allen. *Black Lung: Anatomy of a Public Health Disaster*. Ithaca NY: Cornell University Press, 1998.

Draper, Elaine. *The Company Doctor: Risk, Responsibility, and Corporate Professionalism*. New York: Russell Sage, 2003.

Engels, Friedrich. *The Condition of the Working Class in England*. London: Panther Books, 1969.

Farmer, Paul. *Pathologies of Power: Health, Human Rights and the New War on the Poor*. Berkeley CA: University of California Press, 2005.

Flynn, L. *Studded with Diamonds, Paved with Gold: Miners, Mining Companies and Human Rights in Southern Africa*. London: Bloomsbury, 1992.

Guild, R., Ehrlich, R., Johnston, J. R. and Ross, M. *A Handbook on Occupational Health Practice in the South African Mining Industry*. SIMRAC (Safety in Mines Research Advisory Committee) Johannesburg, 2001.

Harries, Patrick. *Work, Culture and Identity: Migrant Laborers in Mozambique and South Africa C 1860–1910*. Portsmouth NH: Heinemann, 1994.

Hobsbawm, Eric. *The Age of Extremes: A History of the World, 1914–1991*. New York: Pantheon Books, 1994.

Innes, Duncan. *Anglo American and the Rise of Modern South Africa*. London: Monthly Review Press, 1984.

International Labour Office. *Workmen's Compensation for Silicosis in the Union of South Africa, Great Britain and Germany*. International Labour Office Studies and Reports, Series F (Industrial Hygiene) No.16. London: P. S. King, 1937.

International Labour Organisation. *Silicosis: Records of the International Conference held at Johannesburg 13–27th August 1930*. London: ILO, 1930.

Jeeves, Alan H. *Migrant Labour in South Africa's Mining Economy: The Struggle for the Gold Mines' Labour Supply 1890–1920*. Kingston Ont. and Montreal, McGill-Queen's University Press, 1985.

Johnstone, Frederick A. *Class, Race and Gold: A Study of Class Relations and Racial Discrimination in South Africa*. London: Routledge & Kegan Paul, 1976.

Judkins, Bennett M. *We Offer Ourselves as Evidence: Toward Workers Control of Occupational Health*. New York: Greenwood Press, 1986.

Katz, Elaine. *The White Death: Silicosis on the Witwatersrand Gold Mines 1886–1910*. Johannesburg: Witwatersrand University Press, 1994.

King, Rina. 'Silicosis in South African Gold Mines: A Study of Risk of Disease for Black Mineworkers'. TAG/WITS Sociology Research, 1985.

Kraak, Gerald. *Breaking the Chains: Labour in South Africa in the 1970s and 1980s.* London: Pluto Press, 1993.

Lanza, Anthony. *Silicosis and Asbestosis.* London: Oxford University Press, 1938.

Lipton, Merle. *Capitalism and Apartheid: South Africa, 1910–1984.* London: Maurice Temple Smith; Aldershot: Gower, 1985.

Malan, Marais. *The Quest of Health: The South African Institute of Medical Research, 1912–1973.* Johannesburg: Lowry Publishers, 1988.

Markowitz, Gerald and Rosner, David. *Deceit and Denial: The Deadly Politics of Industrial Pollution.* Berkeley CA: University of California Press, 2002.

Martiny, Oluf. 'My Medical Career'. Unpublished manuscript, Johannesburg, September 1999.

Marks, Shula and Rathbone, Richard (eds). *Industrialisation and Social Change in South Africa,: African Class Formation, Culture and Consciousness, 1870–1930.* London: Longman, 1982.

Mavrogordato, A. 'Contributions to the Study of Miners' Phthisis'. Original unpublished Manuscript undated. Adler Medical Museum, University of the Witwatersrand, Johannesburg.

McCrae, John. *The Ash of Silicotic Lungs.* Johannesburg: The South African Institute for Medical Research, 1918[1913].

McCulloch, Jock. *Asbestos Blues: Labour, Capital, Physicians and the State in South Africa.* Oxford: James Currey, 2002.

McCulloch, Jock. *Colonial Psychiatry and 'the African Mind'.* Cambridge: Cambridge University Press, 2005.

McCulloch, Jock and Tweedale, Geoffrey. *Defending the Indefensible: The Global Asbestos Industry.* Oxford: Oxford University Press, 2008.

McIvor, Arthur and Johnston, Ronald. *Miners Lung: A History of Dust Diseases in British Coal Mining.* London: Ashgate, 2007.

Michaels, David. *Doubt is Their Product: How Industry's Assault on Science Threatens Your Health.* Oxford: Oxford University Press, 2008.

Moodie, T. Dunbar. *Going for Gold: Men, Mines and Migration.* Berkeley CA: University of California Press, 1994.

Morrison, Sue. *The Silicosis Experience in Scotland: Causality, Recognition and the Impact of Legislation during the Twentieth Century.* Berlin; Lambert Academic, 2010.

Murray, Colin. *Families Divided: The Impact of Migrant Labour in Lesotho.* Cambridge: Cambridge University Press, 1981.

Nash, June. *We Eat the Mines and the Mines Eat Us: Dependency and Exploitation in Bolivian Tin Mines.* New York: Columbia University Press, 1993.

Orenstein, A. J. (ed.). *Proceedings of the Pneumoconiosis Conference held at the University of the Witwatersrand, Johannesburg, 9–24 February 1959.* London: J & A Churchill, 1960.

Oreskes, Naomi and Conway, Erik M. *Merchants of Doubt.* New York: Bloomsbury Press, 2010.

Packard, Randall, M. *White Plague, Black Labor: Tuberculosis and the Political Economy of Health and Disease in South Africa.* Berkeley CA: University of California Press, 1989.

Pallister, David, Stewart, Sarah and Lepper, Ian. *South Africa Inc.: The Oppenheimer Empire.* New Haven CT: Yale University Press, 1987.

Patterson, H. S. *Dust.* Johannesburg: Transvaal Chamber of Mines, 1936.

Phimister, Ian and Van Onselen, Charles. *Studies in the History of African Mine Labour in Colonial Zimbabwe.* Gwelo: Mambo Press, 1978.

Posel, Deborah. *The Making of Apartheid 1948–1961: Conflict and Compromise.* New York: Oxford University Press; Oxford: Clarendon Press, 1997.

Roberts, Jaine, *The Hidden Epidemic Amongst Former Miners: Silicosis, Tuberculosis and the Occupational Diseases in Mines and Works Act in the Eastern Cape, South Africa.* Westville S.A.: Health Systems Trust, June 2009.

Rosner, D. and Markowitz, G. *Deadly Dust: Silicosis and the Politics of Occupational Disease in Twentieth-Century America,* 2nd edn. Ann Arbor MN: Michigan University Press, 2006.

Shapiro, H. A. (ed.). *Pneumoconiosis: Proceedings of the International Conference, Johannesburg 1969.* Cape Town: Oxford University Press, 1970.

Simons, H. J. *Migratory Labour, Migratory Microbes. Occupational Health in the South African Mining Industry: The Formative Years 1870–1956.* Unpublished manuscript, 1960.

Tropp, Jacob A. *Natures of Colonial Change: Environmental Relation in the Making of the Transkei.* Athens OH: Ohio University Press, 2006.

Tuberculosis Research Committee. *Tuberculosis in South African Natives with Special Reference to the Disease Amongst the Mine Labourers on the Witwatersrand.* Publication no. 30. Johannesburg: South African Institute for Medical Research, 1932.

Turrell, Robert *Capital and Labour on the Kimberley Diamond Fields 1871–1890.* Cambridge: Cambridge University Press, 1987.

Tweedale, Geoffrey. *Magic Mineral to Killer Dust: Turner & Newall and the Asbestos Hazard.* Oxford: Oxford University Press, 2000.

Van Onselen, Charles. *Chibaro: African Mine Labour in Southern Rhodesia.* Johannesburg: Ravan Press, 1980.

Watt, Andrew. *History of Miners' Phthisis on the Rand from 1903 to 1916.* Unpublished manuscript May 1925. Adler Medical Museum, University of the Witwatersrand.

Watt, Andrew, Irvine, L. G., Johnson, I and Steuart, W. *Silicosis (Miners' Phthisis) on the Witwatersrand.* Pretoria: Government Printer, 1916.

Weindling, P. (ed.). *The Social History of Occupational Health.* London: Croom Helm, 1985.

Wikeley, N. J. *Compensation for Industrial Disease.* Aldershot: Dartmouth Publishing, 1993.

Wilmot, James G. *Our Precious Metal: African Labour in South Africa's Gold Industry, 1970–1990.* Cape Town: David Philip, 1992.

Wilson, Francis. *Labour in the South African Gold Mines 1911–1969.* Cambridge: Cambridge University Press, 1972.

Wilson, Francis. *Migrant Labour in South Africa.* Johannesburg: The South African Council of Churches, 1972.

Wilson, Francis and Ramphele, Mamphela. *Uprooting Poverty: The South African Challenge.* Report for the Second Carnegie Inquiry into Poverty and Development in South Africa. New York: W. W. Norton, 1989.

Wylie, Diana. *Starving on a Full Stomach: Hunger and the Triumph of Cultural Racism in Modern South Africa.* Charlottesville VA and London: University of Virginia Press, 2001.

Yudelman, David. *The Emergence of Modern South Africa: State, Capital, and the Incorporation of Organised Labour on the South African Gold Fields, 1902–1939.* Cape Town, David Philip, 1984.

Articles

Beadle, D. G. "An Epidemiological Study of the Relationship between the Amount of Dust Breathed and the Incidence of Silicosis in South African Gold Miners" in *Proceedings of the Mine Medical officers Association* Vol. XLV July–September 1965 391: 31–37.

Beadle, D. G. "Recent Progress in Dust Control in South African Gold Mines" in H. A. Shapiro (ed.), *Proceedings of the International Conference on Pneumoconiosis 1969.* Johannesburg: Oxford University Press, 1970: 69–78.

Becklake, Margaret, du Preez, L. and Lutz, W. "Lung Function in Silicosis of the Witwatersrand Goldminer" *American Review of Tuberculosis and Pulmonary Disease* 77 (3), March 1958: 400–12.

Bowden, B. and Penrose, B. "Dust, Contractors, Politics and Silicosis: Conflicting Narratives and the Queensland Royal Commission into Miners' Phthisis, 1911" *Australian Historical Studies* 37, (128), 2006: 89–107.

Bufton, M. W. and Melling, J. "A Mere Matter of Rock: Organized Labour, Scientific Evidence and British Government Schemes for Compensation of Silicosis and Pneumoconiosis among Coalminers, 1926–1940" *Medical History* 49, (2), 2005:155–78.

Burke, Gillian and Richardson, Peter. "The Profits of Death: A Comparative Study of Miners' Phthisis in Cornwall and the Transvaal, 1876–1918" *Journal of Southern African Studies*, 4 (2), 1978: 141–71.

Cowie, R. L. "The Modern Treatment of Tuberculosis" in *Proceedings of the Mine Medical Officers' Association of South Africa January–August, 1978* LVII (425): 56–57.

Crush, Jonathan , Ulicki, Theresa, Tseane, Teke and Van Veuren, Elizabeth. "Undermining Labour: The Rise of Sub-contracting in South African

Gold Mines" *Journal of Southern African Studies* 27 (1), 2001: 5–31.

Dangerfield, L. F. "Pulmonary Tuberculosis in South Africa and the Problem of the Native Mine Labourer" in *Proceedings of the Transvaal Mine Medical Officers' Association, March 1943* XXII (249): 171–79.

Daubenton, F. "Training and Specialisation of Mine Medical Officers" in *Proceedings of the Transvaal Mine Medical Officers' Association, May 1935* XV (4): 75–78.

Fetter, Bruce. "Changing Determinants of African Mineworker Mortality: Witwatersrand and the Copperbelt, 1911–1940" *Civilisations* 41, 1993: 347–59.

Gandevia, Bryan. "The Australian Contribution to the History of the Pneumoconioses" *Medical History* 17 (4), 1973: 368–79.

Girdler-Brown, B. White, N. V, Ehrlich, R. I. and Churchyard, G. "The Burden of Silicosis, Pulmonary Tuberculosis and COPD Among Former Basotho Goldminers" *American Journal of Industrial Medicine* 51 (9), 2008: 640–47.

Glass, Yette. "The African in the Mining Industry" in *1921–1971 Mine Medical Officers' Association of South Africa. Proceedings of the Jubilee Congress of the Association*, 23 to 25, March 1971: 6–15.

Hall, Arthur J. "Some Impressions of the International Conference on Silicosis" *The Lancet* 216 (5586), 1930: 657–58.

Hnizdo, E. and Murray, J. "Risk of Pulmonary Tuberculosis Relative to Silicosis and Exposure to Silica Dust in South African Gold Miners" *Occupational and Environmental Medicine* 55 (7), 1998: 496–502.

Irvine, L. G. and Macaulay, D. "The Life-history of the Native Mine Labourer in the Transvaal" *Journal of Hygiene* 6 (2) 1906: 149–174.

Jeeves, Alan H. "William Gemmill and South African Expansion, 1920–1950" *The Making of Class*, History Workshop, University of the Witwatersrand, 14 February 1987.

Kippen, S. "The Social and Political Meaning of the Silent Epidemic of Miners' Phthisis, Bendigo 1860–1960" *Social Science and Medicine* 41 (4), 1995: 491–99.

Louw, J. A. "The Miniature Chest Radiograph: Its Uses, Value and Problems" *in Proceedings of the Mine Medical Officers' Association* XLV (392) 1965: 80–84.

MacColl, N. R. A. "The Early Diagnosis of Tuberculosis" in *Proceedings of the Transvaal Mine Medical Officers' Association* XX (218), 1940: 1–2.

Marks, Shula. "The Silent Scourge? Silicosis, Respiratory Disease and Gold-Mining in South Africa" *Journal of Ethnic and Migration Studies* 32 (4), 2006: 569–89.

Martiny, O. "Socio-Medical Problems in the Mining Industry in Relation to Altered Recruiting" in *Proceedings of the Mine Medical Officers' Association of S.A. May 1979–April 1980* LVIII (427): 8–13.

Martiny, O. "A Tuberculosis Programme for South Africa, the Mines and Voluntary Agencies" in *Proceedings of the Mine Medical Officers' Association of S.A. May–September 1970* XLIX (407):162–67.

Mavrogordato, A. "Contributions to the Study of Miners' Phthisis" *Occasional Publications*, South African Institute of Medical Research 3 (19), 1926.

McCulloch, Jock. "Asbestos, Lies and the State: Occupational Disease and South African Science" *African Studies* 64 (2), 2005: 201–16.

McCulloch, Jock "Beating the Odds: The Quest for Justice by South African Asbestos Mining Communities". *The Review of African Political Economy* 32 (103), 2005: 63–77.

McCulloch, Jock. "Hiding a Pandemic: Dr G. W. H. Schepers and the Politics of Silicosis in South Africa" *The Journal of Southern African Studies* 35 (4), 2009: 835–48.

McCulloch, Jock. "Saving the Asbestos Industry, 1960 to 2003" *Public Health Reports* 121 (5), 2006: 609–14.

Millar, A. "Brief Notes on the Prevalence of Tuberculosis in South Africa with Special Reference to Native Mine Labourers". In *Proceedings of the Transvaal Mine Medical Officers' Association March 1943* XXII (249): 167–70.

Mills, C. "The Emergence of Statutory Hygiene Precautions in the British Mining Industries, 1890–1914" *Historical Journal* 51 (1), 2008: 145–68.

Murray, Jill, Davies, Tony and Rees, David. "Occupational Lung Disease in the South African Mining Industry: Research and Policy Implementation" *Journal of Public Health Reports* Vol. 32 (S1), 2011: 65–79.

Packard, Randall M. "Tuberculosis and the Development of Industrial Health Policies on the Witwatersrand, 1902–1932" *Journal of Southern African Studies* 13 (2), 1987: 187–209.

Packard, Randall M. "The Invention of the Tropical Worker: Medical research and the Quest for Central African Labor on the South African Gold Mines, 1903–36" *Journal of African History* 34 (2), 1993: 271–92.

Penrose, Beris "Medical Monitoring and Silicosis in Metal Miners: 1910–1940", *Labour History Review* 69 (3), 2004: 285–303.

Rees, David, Murray, Jill, Nelson, Gill and Sonnenberg, Pam. "Oscillating Migration and the Epidemics of Silicosis, Tuberculosis, and HIV Infection in South African Gold Miners" *American Journal of Industrial Medicine* 53 (4), 2010: 398–404.

Retief, F. "The 'Clinical Side' of Tuberculosis" in *Proceedings of the Transvaal Mine Medical Officers' Association* XIX (215), February 1940: 237–39.

Rosner, D., Markowitz, G. "Workers, Industry, and the Control of Information: Silicosis and the Industrial Hygiene Foundation" *Journal of Public Health Policy* 16 (1), 1995: 29–58.

Silvestri, S. "Change in the World of Occupational Health: Silica Control Then and Now" *Journal of Public Health Policy* 26 (2), 2005: 203–05.

Skaife, W. F. "Detection and Prevention of Tuberculosis" in *Proceedings of the Transvaal Mine Medical Officers' Association* V (5), 1925: 1–4.

Smith, A. "Weighing" in *Proceedings of the Transvaal Mine Medical Officers' Association* V (5) 1925: 4–5.

Steen, T. W, Gyi, K. M. and White, N. W. et al. "Prevalence of Occupational Lung Diseases among Botswana men formerly employed in the South African Mining Industry" *Occupational and Environmental Medicine* No. 54 (1), 1997:19–26.

Stuckler, David, Basu, Sanjay, McKee, Martin and Lurie, Mark. "Mining and Risk of Tuberculosis in Sub-Saharan Africa" *American Journal of Public Health* 101 (3), 2010: 524–30.

Tweedale, Geoffrey and Flynn, Laurie. "Piercing the Corporate Veil: Cape Industries and Multinational Corporate Liability for a Toxic Hazard, 1950–2004". *Enterprise and Society* 8 (2), 2007: 268–96.

Vergara, A. "The Recognition of Silicosis: Labor Unions and Physicians in the Chilean Copper Industry, 1930s–1960s" *Bulletin of the History of Medicine* 79 (4), 2005: 723–48.

Watkins-Pitchford, W. "Miners' Phthisis: Its Cause, Nature, Incidence and Prevention" in The Pan-Pacific Science Congress Melbourne Session, *The Medical Journal of Australia* 11 (13), 1923: 325–27.

Watkins-Pitchford, W. "The Diagnosis of Silicosis" *Medical Journal of Australia* 2 (15) 1923: 382–84.

Watkins-Pitchford, W. "The Silicosis of the South African Gold Mines, and the Changes Produced in it by Legislative and Administrative Effort" *Journal of Industrial Hygiene* 9 (4), 1927: 129–30.

Watt, Andrew. "History of Miners' Phthisis on the Rand from 1903 to 1916". Unpublished manuscript, Adler Medical Museum, University of the Witwatersrand, May 1925.

White, Neil. "Is the ODMW Act Fair? A Comparison of the Occupational Diseases in Mines and Works Amendment Act, 1993 and the Compensation of Occupational Injuries and Diseases Act, 1993 with respect to compensation of Pneumoconiosis". Unpublished paper, July 2004.

Wilson, Francis. "Minerals and Migrants: How the Mining Industry Has Shaped South Africa" *Daedalus* 130 (1), 2001: 99–121.

Yudelman, David and Jeeves, Alan. "New Labour Frontiers for Old: Black Migrants to the South African Gold Mines, 1920–85" *Journal of Southern African Studies* 13(1), 1986: 101–24.

Zwi, Anthony, Fonn, Sharon and Steinberg, Malcolm. "Occupational Health and Safety in South Africa: The Perspectives of Capital, State and Unions" *Social Science and Medicine* 27 (7), 1988: 691–702.

Theses

Beadle, D. G. *The Performance and Improvement of Dust Sampling Instruments.* Masters of Science thesis, Rhodes University, 1954.

Cowie, R. L. *Pulmonary Tuberculosis in South African Gold Mines: Determinants of Relapse after Treatment.* Masters of Science thesis, McGill University, 1987.

Cowie, R. L. *Silicosis, Pulmonary Dysfunction and Respiratory Symptoms in South African Gold Miners.* M.D. thesis, Faculty of Medicine, University of Cape Town, 1987.

Donsky, Isidore. *A History of Silicosis on the Witwatersrand Gold Mines, 1910–1946.* Ph.D. thesis, Rand Afrikaans University, November 1993.

Roberts, J. E. *What is the Price of 80 KGs? The Failure of the Detection of, and Compensation for Asbestos-related Disease: Social Exclusion in Sekhukhuneland.* M.A. thesis, University of Natal, December 2000.

Slade, G. F. *The Incidence of Respiratory Disability in Workers Employed in Asbestos Mining with Special Reference to the Type of Disability Caused by the Inhalation of Asbestos Dust.* M.D. thesis, Witwatersrand University, 1930.

Smith, Matthew John. *Working in the Grave: the Development of a Health and Safety System on the Witwatersrand Gold Mines, 1900–1939.* M.A. thesis, Rhodes University, 1993.

Trapido, Anna. *An Analysis of the Burden of Occupational Lung Disease in a Random Sample of Former Gold Mineworkers in the Libode District of the Eastern Cape.* Ph.D. thesis, University of the Witwatersrand, 2000.

INDEX